AF389387

Toxoplasmosis: Epidemiology, Prevention and Control

Toxoplasmosis: Epidemiology, Prevention and Control

Ionela Hotea
Gheorghe Dărăbuș

Basel • Beijing • Wuhan • Barcelona • Belgrade • Novi Sad • Cluj • Manchester

Authors

Ionela Hotea
Faculty of Veterinary Medicine,
University of Life Sciences "King Mihai I"
Timisoara, Romania

Gheorghe Dărăbuș
Faculty of Veterinary Medicine,
University of Life Sciences "King Mihai I"
Timisoara, Romania

Editorial Office
MDPI
St. Alban-Anlage 66
4052 Basel, Switzerland

For citation purposes, cite as indicated below:

Lastname, Firstname, Firstname Lastname, and Firstname Lastname. Year. *Book Title*. Series Title (optional). Basel: MDPI, Page Range.

ISBN 978-3-0365-2699-7 (Hbk)
ISBN 978-3-0365-2698-0 (PDF)
doi.org/10.3390/books978-3-0365-2698-0

Cover image courtesy of Ionela Hotea.
This book was published using funds from the University of Life Sciences "King Mihai I", Timisoara, Romania, and the Research Institute for Biosecurity and Bioengineering, Timisoara, Romania.

Contents

About the Authors

Ionela Hotea graduated the Faculty of Veterinary Medicine from Timisoara, Romania in 2007. In the same institution, she followed PhD studies in the field of Veterinary Parasitology, graduated in 2011. With more than 15 years of experience in the field of veterinary medical scientific research, she has contributed to the implementation of more than 10 research projects in the field. She has published more than 170 national and international scientific papers and 8 books in the field of Life Sciences. She won the national award "Eminent Researcher"—University Horizons Association—in 2017 and awards for prestigious scientific publications. She is a member of the Project Management Unit of the University, Member of the Editorial Board and Scientific Committee of the journal "Advance Research in Life Sciences" USV Timisoara, as well as in various nationally and internationally recognized scientific associations. She is currently a Lecturer at the Faculty of Veterinary Medicine of the University of Life Sciences "King Mihai I" from Timisoara, Romania, where, in addition to her teaching activity, she passionately continues her research activities approaching various scientific topics in the veterinary medical field.

Professor Emeritus Gheorghe Dărăbuș, DVM, PhD, Dr.h.c. is a professor of Parasitology and parasitic diseases, Faculty of Veterinary Medicine at the University of Life Sciences "King Michael I" from Timișoara, Romania. He is a full member of the Academy of Agricultural and Forestry Sciences "Gheorghe Ionescu Sisesti" (ASAS) and vice-president of the Veterinary Medicine Section of this academy. He has published more than 500 scientific works in the country and abroad. His works refer to studies on descriptive and molecular epidemiology, pathogenesis, diagnosis and parasitological control of diseases such as toxoplasmosis, cryptosporidiosis, chicken eimeria, cystic echinococcosis, trichostrongylidosis, etc. He is the author and/or co-author of four patents. He is the first author or collaborator of 27 books. As a reward, he received two "Traian Săvulescu" awards from the Romanian Academy (2012 and 2016) and two ASAS awards, namely the "Ion Adameșteanu" award (2011) and the "Paul Riegler" award (2015). He has 8 UEFISCDI awards. He received the "Diploma und Medaille zur 200 jahrfeier 1790–1990—Ludwig-Maximilians" (2009), University of Munich. He has several dozen awards and gold medals, obtained together with the discipline's collective, at the national and international salons of invention. He was and is a member of editorial teams or scientific committees of some scientific journals and events, organizer of scientific events, reviewer for journals and national and international scientific events.

Preface

Approaching a single disease, a very important zoonosis, has the advantage that the exhaustive knowledge gained in the research of this disease can be exploited. We believe that this monograph is timely, on the one hand, because there are still many unknown things about this zoonosis. On the other hand, it is necessary to make known the concerns of Romanian researchers related to this protozoonosis.

In the current work, which I have titled "Toxoplasmosis: Epidemiology, Prevention and Control", we propose to synthesize in one volume both the general knowledge regarding the biology, epidemiology and diagnosis of toxoplasmosis, but also some recommendations and elements of parasitological control. The book has as its starting point a doctoral thesis developed by Dr. Ionela Hotea, coordinated by Prof. Dr. Gheorghe Dărăbuș. In addition to worldwide knowledge, the paper gathers the Romanian contributions to the study of this protozoonosis, especially those brought by the parasitology school at the Faculty of Veterinary Medicine in Timisoara.

The book is written in a conventional system, offering human doctors, veterinarians and students, in a single volume, classic knowledge but also news about the evolution of this parasitosis, the achievements and progress of Romanian parasitology, supplemented with information from the world specialized literature. We hope that this book will contribute to raising the level of knowledge of human and veterinary doctors, to optimizing the training of students in the field, to the development of the training process in university education, to the development of the continuous learning process. This volume may also be a starting point for the development of Romanian scientific research.

Being the first edition of this book, some inadvertences or mistakes may be inherent. Therefore, we ask the readers to bring to our attention any observations to consider them in a new edition.

Gheorghe Dărăbuș
University of Life Sciences "King Mihai I" from Timisoara
Timis, Romania

1. Introduction

Toxoplasmosis is one of the most common parasites in humans and animals, ranking third in the world's zoonosis. Toxoplasmosis causes economic losses around livestock, through abortions, embryonic mortality or non-viable products. The intermediate hosts are numerous species of mammals and birds, and also humans. In felids, the definitive hosts, *Toxoplasma gondii* develops in the small intestine, while in intermediate hosts, the parasite is localized in different parts of the body, with the formation of pseudocysts and then cysts in the brain, muscles and various organs (Tenter et al. 2000; Petersen et al. 2006). At the European level, the interest of specialists in toxoplasmosis is remarkable (Cermáková et al. 2005; Colville and Berryhill 2007; Durecko et al. 2004). In Romania, research related to this zoonosis is under development.

Toxoplasmosis is the subject of numerous bibliographic studies conducted worldwide, with equal reference to humans and animals (Tenter et al. 2000; Dubey 2004). Serological surveillance has shown that *T. gondii* infection is widespread in farm animals. The most receptive species are sheep, goats and pigs, then leporids, horses, cattle and also birds. In sheep, the prevalence of toxoplasmosis varies between 3% in Pakistan and 95.7% in Turkey, and in goats between 3.2% in Mexico and 90.9% in the Netherlands. In pigs, the prevalence of *T. gondii* infection is between 0.74% in Ontario, Canada and 90.4% in Minas Gerais, Brazil. Various prevalence has also been identified in less susceptible species. Thus, in horses, the prevalence of toxoplasmosis varies between 1% in Sweden and 48.1% in Egypt, and in cattle between 1.6% in Cordoba, Spain and 91.8% in Lombardy, Italy. In buffaloes, the prevalence of toxoplasmosis is very low, with the highest value (8.8%) being identified in Iran. Viable forms of the parasite could not be isolated from this species. Buffaloes are considered unresponsive to this disease. Among pets, in cats, the prevalence was reported between 2.1% in Guangdong, China, and 89.3% in Colombia, and in dogs, between 3.7% in the Czech Republic and 88.5% in Mato Grosso, Brazil (Dubey 2010). In wild species from zoos, the prevalence of *T. gondii* infection is varied. Toxoplasmosis can be found in camels (0–26.3%), monkeys (10.2–30.2%), marsupials (3.3–26.1%), wild carnivores (0–100%), bears (35–70%), deer (1.2–90%) and also in wild boar (1–83.7%) and hares (6.1–26.9%) (Dubey 2010).

There are many factors that influence epidemiology, such as the management of farm, breeding and maintenance; cat density; climatic conditions that influence the growth of oocysts in the environment (temperature, humidity, wind); geographical area; and different culinary and cultural habits of consumers (Van der Giessen et al. 2007; de Moura et al. 2007).

The prevalence of toxoplasmosis varies depending on the living conditions of the animals, being higher in free-range cats, which have the opportunity to hunt small mammals or birds. It is much lower in domestic-raised animals without access to the outside. In farm animals, the prevalence is higher where animals have direct or indirect contact, through feed, with cats. The seroprevalence of *T. gondii* infection differs from country to country, but also within the same country, from one region to another, or even within the same city. These differences depend on the degree of civilization and culture of the human population. Controlled feeding of cats with cooked food or commercial food can greatly reduce the incidence of this disease.

Instead, the lack of food forces these predators to hunt in order to survive and thus, the biological cycle of the parasite continues. Additionally, supervised feeding and providing heat-treated food, where appropriate, of farm animals can significantly reduce the incidence of toxoplasmosis.

2. Toxoplasmosis Etiology and History

Toxoplasmosis is a general protozooses produced by coccidia belonging to the family *Toxoplasmidae*. It parasitizes in various organs and tissues in over 350 species of vertebrates. From this point of view, it is perhaps the most widespread protozoan (Garcia et al. 2005; Dărăbuş et al. 2006; Ghazy et al. 2007).

Toxoplasma gondii is considered one of the most complex parasites known to date. It has a heterogeneous biological cycle and can infect all species of warm-blooded animals (mammals and birds), including humans. The parasite can be found anywhere in the world, which demonstrates its medical and veterinary importance, as it can cause abortion and birth defects in intermediate hosts. Due to its zoonotic importance, *T. gondii* is the most studied of the coccidia.

The first detailed description of tachyzoites (asexual forms) in the spleen, liver and blood of some rodent species in North Africa, called *Ctenodactylus gondii*, was made by Nicolle and Manceaux in 1908 (cit. by Tenter et al. 2000). They were also the ones who introduced the genus *Toxoplasma*, and *T. gondii* became the type species of this genus (Table 1):

Table 1. Taxonomic classification (by Dubey and Lappin 1998).

Kingdom	Protista
Subkingdom	Protozoa
Phylum	Apicomplexa
Class	Sporozoasida
Order	Eucoccidiorida
Family	Sarcocystidae
Genus	Toxoplasma
Species	*gondii*

Source: Authors' compilation based on data from Dubey and Lappin 1998.

During the first half of the 19th century, Toxoplasma species were named after the host species in which they were observed. In the late 1930s, biological and immunological studies showed that all other species discovered were, in fact, identical to *T. gondii* (Table 2) (Tenter et al. 2000).

The discovery of the coccidian nature in *Toxoplasma gondii* was made in the 1960s, when detailed studies revealed ultrastructural similarities between *T. gondii* merozoites and Eimeria species merozoites. Highlighting infesting forms for intermediate hosts, present in cat feces, and later, description of sexual stages in the cat's small intestine elucidated the coccidian and heterogeneous nature of the biological cycle for *T. gondii* (Dubey et al. 2006a).

Table 2. Historical data about *Toxoplasma gondii* (by Tenter et al. 2000).

Year	The Event	Year	The Event
1900	Description of *Toxoplasma* as a parasite in Java sparrows.	1952	Description of the classic tetrade of symptoms in congenital toxoplasmosis in humans (chorioretinitis, cerebral calcifications, hydrocephalus or microcephaly, psychomotor disorders).
1908	First description of *Toxoplasma* as a human tissue cyst (associated with sarcosporidiosis).	1953–1954	The first case of toxoplasmic encephalitis in a patient with Hodgkin's disease.
1908	Description of *T. gondii* merozoites in gondii mice by Nicolle and Manceau.	1957	Toxoplasmic abortion is recognized in sheep.
1908	Discovery in rabbits (*Toxoplasma cuniculi*).	1954–1956	Elaboration of the hypothesis that the horizontal transmission to humans occurs through slightly cooked pork.
1909	Introduction of the genus *Toxoplasma* (with type species: *T. gondii*).	1959	Repeated congenital transmission in mice.
1910	Description of the disease in a domestic animal (dog).	1959	Serological evidence of toxoplasmosis in vegetarians.
1923	First case of congenital toxoplasmosis in an 11-month-old child with hydrocephalus and microphthalmia.	1960	Discovery that tissue cysts are resistant to proteolytic enzymes.
1928	The first description of tissue cysts as a persistent stage in the intermediate host.	1960	Discovery of the major sequelae left by congenital toxoplasmosis in humans.
1937	Demonstration of congenital transmission.	1965	Recognition of the coccidian nature of *T. gondii* based on highlighting the ultastructure of extraintestinal merozoites.

Table 2. *Cont.*

Year	The Event	Year	The Event
1937	The first fatal case of disseminated toxoplasmosis in a 22-year-old adult.	1965	Epidemiological evidence that horizontal transmission to humans occurs through poorly cooked meat.
1937–1939	Recognition of *T. gondii* as a causative agent of encephalomyelitis in newborns.	1965	The hypothesis that an infectious stage of *T. gondii* is passed into the environment with cat feces.
1939	Description of the classic triad of symptoms in congenital toxoplasmosis in humans (chorioretinitis, hydrocephalus, encephalitis followed by cerebral calcification).	1968	Recognition of *T. gondii* as a complication in patients with malignancies.
1939	Highlighting the similarities between human and animal lesions based on biological and immunological analogies.	1969	Identification of *T. gondii* oocysts.
1940–1941	Recognition of *T. gondii* as a causative agent of acute form in adults	1970	Description of the sexual phase of the biological cycle in the small intestine in cats
1941–1942	Detailed description of toxoplasmic encephalitis in children.	1970–1972	Establishment of final and intermediate hosts.
1942	Recognition of vertical transmission in humans.	1969–1972	Recognition of the epidemiological role of cats in the spread of *T. gondii* in different geographical areas (including isolated islands).
1948	Introduction of the methylene blue test for *T. gondii* antibodies.	1972	Description of 5 types of *T. gondii* from the intestinal epithelium of cats.
1951–1952	Recognition of *T. gondii* as a causative agent of lymphadenopathy in humans.	1981–1982	The first case of CNS toxoplasmosis in AIDS patients.

Table 2. *Cont.*

Year	The Event	Year	The Event
1952	Recognition of *T. gondii* as a causative agent of chorioretinitis in humans.	1984	Recognition of *T. gondii* as an opportunistic pathogen in AIDS patients.
		1995–1999	The most significant epidemic of acute toxoplasmosis in humans (100 individuals aged 6 to 83 years) associated with the presence of oocysts in the municipal drinking water tank.

Source: Authors' compilation based on data from Tenter et al. 2000.

For more than three decades, *T. gondii* has been considered the only valid species of the genus Toxoplasma, but recent studies have shown that there are at least two clone lines in *T. gondii*, one virulent for mice and the other avirulent for mice. A study by Johnson in 1997 suggested that vertical transmission of the virulent line to mice is of greater epidemiological importance than previously thought (Johnson 1997; da Silva et al. 2005; Djurković-Djaković et al. 2005; Speer and Dubey 2005).

Throughout its evolution, *T. gondii* has been shown to have several routes of transmission but, for three decades, no conclusion has been reached on the most important of these. Congenital toxoplasmosis has long been thought to be the most important—the result of vertical transmission during pregnancy. On the other hand, some information is known about the horizontal transmission of *T. gondii* between different host species, about the large natural reservoir of *T. gondii* or about the epidemiological impact of various sources that cause infections or diseases in humans (Dărăbuş et al. 2006).

In the final host—a feline—toxoplasmosis usually develops asymptomatically, while in the intermediate hosts, it can cause abortions or even death. Economic losses are significant due to reduced birth rates and abortions in farm animals, especially sheep and pigs. Humans are also receptive to *T. gondii*, and can cause abortion, stillbirth, birth defects, neurological signs, ocular symptoms, etc. (Negash et al. 2004; Brandão et al. 2006; Gauss et al. 2006; Hill et al. 2006; Jones et al. 2007).

3. Biology and Morphology of *Toxoplasma gondii*

3.1. Lyfe Cicle

T. gondii is a ubiquitous parasite that can be found anywhere in the world and can infect even the most unusual species of animals, as well as many types of host cells. The biological cycle of *T. gondii* is optionally heterogeneous. Intermediate hosts are probably all warm-blooded species, including most farm animal species as well as humans. Definitive hosts are represented by members of the feline family, especially the domestic cat (Johnson 1997; Garcia et al. 2005; Dangolla et al. 2006; Figueroa-Castillo et al. 2006; Mucker et al. 2006; Ryser-Degiorgis et al. 2006; Ghazy et al. 2007).

Until 1970, only the asexual stages—tachyzoites and bradyzoites—were known. The sexual forms of the cycle and the presence of resistant forms in the environment—i.e., oocysts—have been explained since 1970. In the last 40 years, several discoveries have been made regarding the complete biological cycle of the parasite, especially in the sexual phase of the cycle, and the impact of these data on the disease and its control.

The transmission of the parasite was a mystery until its discovery in the rodent Ctenodactylus gondii. Chatton and Blanc (1917, cit. by Dubey and Jones 2008) reported that this rodent did not become infected in the wild, but in captivity. Ctenodactylus gondii lives in the mountains of southern Tunisia. Rodents were used in research for leishmaniasis at the Pasteur Institute in Tunisia. Chatton and Blanc assumed that *T. gondii* was transmitted through arthropods, the parasite being identified in the host's blood (Dubey and Jones 2008).

In the intermediate host, *T. gondii* has two phases of asexual development. In the first phase, tachyzoites multiply rapidly by repeated endodyogeny in different types of host cells. Pseudocysts with tachyzoites are specific for acute toxoplasmosis. In the hosts that survive this evolution, tachyzoites invade any cell, either passively (by hyaluronidase and lysozyme) or actively (by phagocytosis) (Cosoroabă 2005).

The tachyzoites are crescent shaped, with the nucleus located centrally, having approximately 6×2 μm and multiplying in almost any cell in the body. Tachyzoites do not completely fill the host cell, whose nucleus remains visible (Lachenmaier et al. 2011). Generally, the nucleus is located towards the posterior extremity or in the central area of the cell. *Toxoplasma gondii* can move by sliding, rippling and rotating (Dubey 2005).

Tachyzoite enters the host cell by actively penetrating its membrane. After penetration, the tachyzoite is rounded and surrounded by a parasitophorous vacuole (PV). It has been suggested that this vacuole is produced by both the parasite and the host cell (Toulah et al. 2011).

Inside the host cell, tachyzoite multiplies asexually by repeated endodyogeny. Endodyogeny (endo = inside; dyo = two; genos = birth) is a specialized form of reproduction, in which two offspring are formed inside the parent parasite (Lachenmaier et al. 2011). Tachyzoites continue to divide by endodyogeny until the host cell is full of parasites. They contain numerous granules with trophic reserves that underlie the speed of multiplication (Dubey 2005).

If cell-destruction occurs, they infect other cells, continuing their development (Lachenmaier et al. 2011). Tachyzoites of the last generation initiate the second phase of asexual development, which results in tissue cysts (Toulah et al. 2011). The transition from the pseudocyst to the cyst is the result of the appearance of the state of immunity and of some genetic factors related to the host. In the genesis of immunity, interferon (INFγ) plays an essential role, inhibiting the multiplication of tachyzoites—a necessary condition for the formation of cysts (Tekay and Ozbek 2007). It has not been shown that tachyzoites disappear completely with the formation of cysts. "Waiting tachyzoites" may persist and cause a resurgence of toxoplasmosis (Cosoroabă 2005).

Tissue cysts grow intracellularly and contain numerous bradyzoites that are similar in structure to tachyzoites, but the bradyzoites are shorter and thinner, and the nucleus is located at the back of the parasite. Bradyzoites contain heat-shock proteins with a role in the trapping phenomenon. The host cell is strongly modified, and its nucleus is reduced to a thin marginal band. Biologically, bradyzoites differ from tachyzoites in that they can survive stomach digestion, while tachyzoites are usually destroyed. Tissue cysts appear to be essential in toxoplasmosis, as no strains of *Toxoplasma gondii* without cysts have been identified (Olariu 2003; Dărăbuş et al. 2006).

Tissue cysts have variable sizes 15–60–100 μm and usually take the shape of the cell they parasitize, being separated from it by a thin wall (<0.5 μm), which is elastic and argentophilic. Its isolation from host tissue is the result of an antigen–antibody reaction for the purpose of immune evasion of the parasite. Young cysts can be only 5 μm and contain only two bradyzoites, while older cysts can contain hundreds of organisms (Toulah et al. 2011).

Tissue cysts have an increased affinity for neural and muscular tissue, which are located mainly in the CNS, eyes, as well as in the skeletal and cardiac muscles. However, to a lesser extent, they can also be found in visceral organs such as the lungs, liver and kidneys (Dubey 2005; Wang et al. 2006; Cambrea et al. 2007).

Not all strains of *T. gondii* have the same ability to form cysts. The rapidly cystogenic ones are slightly pathogenic. Strains proven to be pathogenic have poor cystogenesis skills or are even acystogenic (RH Sabin strain) (Cosoroabă 2005).

Tissue cysts represent the terminal phase of the biological cycle in the intermediate host, and from that moment, they begin infesting. In some intermediate host species, they may persist throughout the animal's lifetime, but the mechanism by which they persist for so long is not yet known.

Many researchers believe that these cysts periodically rupture, and bradyzoites turn into tachyzoites that invade other cells and turn back into bradyzoites, forming a new tissue cyst. If these cysts are ingested by a definitive host, they initiate another phase of asexual proliferation that begins with multiplication by endodyogeny, followed by endopolygeny in the epithelial cells of the small intestine. The terminal stage of this asexual multiplication initiates the sexual phase of the biological cycle, the sexual stages also developing in the epithelium of the small intestine (Dubey and Beattie 1990).

The intraepithelial biological cycle is found only in the definitive host. The evolutionary cycle that starts from the ingestion of their cysts with bradyzoites has been precisely studied. Many cats become infected by consuming intermediate hosts that have tissue cysts. Bradyzoites are released from cysts in the cat's stomach or intestine due to the dissolution of the cyst wall by digestive enzymes. Bradyzoites enter the epithelial cells of the small intestine and the development of forms (A → E) that precede the asexual phase are initiated. These forms, A → E, are equivalent to the schizonts of other intestinal coccidia. After an indefinite number of generations, the merozoites developed from the D or E-type forms will determine the formation of micro- or macrogamonts. Microgamontes divide and form biflagellate microgametes, which are released and head for the macrogamonts they penetrate. A wall forms around the fertilized macrogamont and becomes oocyst (Shakespeare 2009). When mature, oocysts are eliminated in the intestinal lumen by rupture of intestinal epithelial cells. Oocysts are unsporulated (non-infesting) when they are excreted in the feces. The sporangium occupies almost the entire inner space of the oocyst, and the sporulation occurs outside the host cat within 1–5 days, depending on the degree of aeration and temperature (Dărăbuş et al. 2006).

The sporulated oocysts are subspherical to elliptical and have a diameter of 11 × 13 μm. Each sporulated oocyst contains two sporocysts. Each sporocyst contains four sporozoites. Sporozoites are banana shaped and can survive in oocysts for several months, even in more precarious environmental conditions (Dubey and Beattie 1990; Speer and Dubey 2005; Cosoroabă 2005).

The entire intraepithelial cycle in *Toxoplasma gondii* is complete 3 to 10 days after ingestion of tissue cysts and occurs in approximately 97% of cats that have never been infested. After ingestion of oocysts or tachyzoites, oocyst formation is delayed for up to 18 days and is found in only 20% of cats that have ingested those forms (Toulah et al. 2011). The differences regarding this delayed cycle and the resistance that some cats have are unclear, but it is assumed that bradyzoites are precursors of intraepithelial replication (Cosoroabă 2005).

In cats, along with the entero-epithelial cycle, in which they have the role of definitive host, another cycle can take place: the extraintestinal cycle, in which the role is of intermediate host. In this case, the bradyzoites enter the lamina propria, where they multiply as tachyzoites, with the possibility of subsequent blood dissemination, in a few hours, in the extraintestinal tissues. Tachyzoites can proliferate by endodyogeny in a number of organs and in a wide variety of cell types, being initially present in large numbers in the lungs and spleen, and after 6–10 days they can invade all organs, including the brain. In immunocompetent mice, the maximum number of lesions and that of tachyzoites occur at approximately 12 days post-infectious, and then, at 21 days, it becomes extremely difficult to identify tachyzoites in organs, even by immunohistochemical (IHC) analysis (Weiss and Kim 2004). *Toxoplasma gondii* persists in the intestinal and extraintestinal tissues of the cat for at least a few months, or even the entire life of the animal. The prepatent period is variable with the infecting element: 3–10 days for tissue cysts (100% infectivity), over 19 days for tachyzoites (50% infectivity) and over 20 days if spore oocysts are ingested (50% infectivity) (Dubey 2005).

T. gondii can be transmitted from the final host to the intermediate one, from the intermediate to the definitive host as well as between the definitive and intermediate hosts. It is not yet known exactly which of these pathways is more important from an epidemiological point of view. However, the prevalence of the disease is not limited by the presence of a particular host species. The biological cycle can continue indefinitely by transmitting their tissue cysts between intermediate hosts (even in the absence of the definitive host) or by transmitting oocysts between the definitive hosts (even in the absence of intermediate hosts) (Johnson 1997).

3.1.1. *Toxoplasma gondii* Reproduction—Endodyogeny

Tachyzoites possess a unique ability: namely the ability to proliferate indefinitely through a distinctive process called endodiogeny, which involves the growth and division of the parasite with the formation of two daughter cells (Toulah et al. 2011). Despite the fact that endodiogeny resembles binary division, it is extremely complex due to the process of formation of the two polarized daughter cells (Lachenmaier et al. 2011). In contrast with the classic patterns of asexual division, observed in other members of the *Apicomplexa* cluster (even in the coccidian stages of *T. gondii*), tachyzoites retain their apical complex until the end of endodiogeny. Although these two processes are simultaneous, mitosis and daughter-cell formation will occur separately (Weiss and Kim 2004).

Cell Duplication—Mitosis

Only a few descriptions are available at the ultrastructural level regarding mitosis in *T. gondii*. These observations can be interpreted by comparison with much more detailed studies on other members of the *Apicomplexa* cluster, especially Eimeria spp. One of the unique aspects of mitosis at the *Apicomplexa* is the preservation of the nuclear membrane intact throughout the division. Coccidia-specific centrioles (150 nm in diameter) consist of nine short tubules (100 nm long), in the middle of which is a central tube. Centrosomes or polar bodies of the spindle are formed by two centrioles oriented in parallel. They are always associated with centrocones (or mitotic spindle poles), usually on the apical face of the nucleus. In the earliest stage of mitosis, a sleeve is formed around the nucleus containing fibrous material. The sleeve corresponds to an invagination of the nuclear envelope, open at both ends to the cytoplasm. The mitotic spindle will polymerize in this sleeve, which will then open into the nucleoplasm through its middle portion, while the poles will give rise to centrocones. Initially, the centrocones are in the form of subspherical invaginations of the nuclear envelope open to the centrosomes, through which the spindle microtubules extend. The intranuclear spindle is usually transient and has been described in a few studies. Most likely, the kinetochores will separate immediately after the release of the sleeve and will assemble on the nucleoplasmic face of the centrocones. Karyokinesis does not depend on mitotic elongation. The centrocones quickly become conical veins of the nuclear envelope, open in the centrosomes, and are covered on the nucleoplasmic face with a multilayered structure corresponding to the kinetochores. The specific feature of this stage is that each mitosis takes place simultaneously with the formation of the two daughter cells (Weiss and Kim 2004).

Tachyzoites Biogenesis

Shortly after the separation of the centrosomes and the formation of the centrocones, the development of the future apical complexes of the daughter cells (in the vicinity of each centrosome) will begin. The details of biogenesis have not been studied as systematically or in as much detail as in the case of *Eimeria* spp. However, the same pattern is known to follow. In the early stage of development, a tension fiber can be observed that originates between the two centrioles and that unites an area where the conoid assembles (Lachenmaier et al. 2011). The inner membrane complex and subpellicular microtubules are initially formed around the conoid, then grow in the posterior direction. This process occurs inside the cytoplasm of the mother cell rather than in association with its plasmalemma, an aspect characteristic of the formation of daughter cells by classical schizogony, carried out in most members of the *Apicomplexa* cluster (Toulah et al. 2011). The early stages have the appearance of a short, flattened cone that then elongates into a cylindrical structure that will eventually surround the mature organism (Weiss and Kim 2004).

As development progresses, the nucleus acquires a "U" shape, and the forming internal complex elongates and captures the nuclei of daughter cells. During this time, cellular organs are also formed (Golgi apparatus, apicoplasts, rhoptries, micronemes) (Weiss and Kim 2004).

As the daughter cells develop, the inner membrane complex of the stem cell ruptures with the organs in the anterior portion (Lachenmaier et al. 2011). The fully developed daughter cells occupy a large part of the cytoplasm of the cell of origin, and their internal membrane complex will come into contact with the stem-cell plasmalemma, resulting in the film (Toulah et al. 2011). Stem-cell cracking begins at the anterior extremity and, before total separation occurs, the daughter cells remain connected by a portion of the residual cytoplasm. Repeated division leads to the accumulation of tachyzoites in parasitophorous vacuoles which, when arranged in flattened fibroblast cells, can acquire the typical rosette appearance (Ferreira-da-Silva et al. 2009).

Endodyogeny is an exclusively asexual form of division that occurs in intermediate hosts (during tachyzoite and bradyzoite formation), which is completely different from the division process that takes place in the final host (with the formation of merozoites) or in oocysts (with the formation of sporozoites) (Weiss and Kim 2004).

3.1.2. Life Cycle in the Definitive Hosts

Coccidian development is limited to the epithelial cells in the small intestine of the definitive host (cat). Regardless of the stage of coccidian development, the parasite is wrapped in a thick-walled parasitophorous vacuole, adjusted accordingly to its size. Electron microscopy showed the stratified arrangement of the vacuolar wall, which in some areas consists of three tightly glued membrane units. It has also been possible to visualize dense, conical structures that protrude through the vacuolar wall and that seem to act as a connecting element between the surface of the parasite and that of the membrane of the parasitophorous vacuole. Unlike parasitophorous vacuoles developed by tachyzoites, those formed in cat intestinal cells are not characterized by the presence of a parietal tubular structure and are not surrounded by mitochondria or the rough endoplasmic reticulum of the host cell. These structural differences correlate with the absence of most proteins that form dense granules. The thick and layered structure of the wall of the parasitophorous vacuole is similar to that observed in certain species of Isospora to which *T. gondii* is related, being totally different from the single-membrane unit found in the genus *Eimeria* (Weiss and Kim 2004).

Asexual Cycle

During coccidian development, only one process of asexual division can be observed—designated as endopolygeny (internal budding)—which has unique structural aspects.

The mentioned process involves the growth of the parasite and the repeated nuclear division, with the formation of an eccentric intranuclear spindle, similar to the one described in endodyogeny. In addition, an increase in the size of mitochondria (which are located mainly at the periphery of the parasite) and the multiple presence of apicoplasts (delimited by four membranes and located centrally) were also observed. The number of nuclear divisions varies depending on the parasite, having direct effects on the number of resulting daughter cells. The presence of this proliferative phase, before the formation of daughter cells, differentiates endopolygeny from endodyogeny. It is not known how to control the number of nuclear divisions but, given that there are marked variations between the parasitic elements, it does not seem to be conditioned by the size of the parasite or a predefined number of divisions (Weiss and Kim 2004).

The element that triggers the completion of the proliferation phase and that initiates the differentiation phase (daughter cell formation) is not yet known. The formation of daughter cells can occur at any time, during or immediately after the final nuclear division. The first evidence of the initiation of daughter-cell formation is the appearance of a conical structure consisting of several flattened vesicles, each with subadjacent longitudinal microtubules, and possessing a conoid at the apex (Toulah et al. 2011).

The mechanism of daughter-cell development is similar to that observed in the endodyogeny of tachyzoites or bradyzoites (Lachenmaier et al. 2011). Due to the fact that it occurs in several nucleolar directions, the term endopolygeny is more appropriate. The simultaneous formation of several daughter cells requires an extremely well-coordinated process of assigning the entire organ equipment to maintain viability. The final stage of development is represented by the invagination of the stem cell plasmalemma, starting from the apical extremity of the daughter cells and progressing posteriorly, the result being the formation of the outer membrane of the film belonging to each daughter cell (Lachenmaier et al. 2011). Often, merozoites remain attached to the posterior extremity by a small amount of residual cytoplasm. These daughter cells are banana shaped and can form fan-shaped structures. Merozoites are released from the host cell into the intestinal lumen, where they can invade enterocytes (Weiss and Kim 2004).

Unlike many species of *Eimeria*, in *T. gondii* there do not appear to be several distinct generations of schizogony, which differ in size and the number of daughter cells formed. However, studies targeting early stages of infection (first 1–3 days) have identified additional asexual processes (Speer and Dubey 2005). It has been reported that some developing parasites have a similar host–parasite relationship and develop endodyogeny and repeated endodyogeny (type B schizonts), and others appear to be an intermediate category (type C schizonts). Type B schizonts have similarities to parasitic elements that invade the small intestine of intermediate hosts. These stages are rare and may be an example in which invasive bradyzoites fail to convert to coccidian development and turn into tachyzoites, similar to the situation observed in intermediate hosts (Toulah et al. 2011). It is also known that the cat can be both a definitive and an intermediate host for *T. gondii*.

Sexual Cycle

After an unknown number of asexual cycles, certain merozoites enter enterocytes and transform into male (microgametocytes) and female (macrogametocytes) gametocytes. In the case of microgametogony, several microgametes are formed (15–30), and macrogametogeny results in the formation of a single female gamete (macrogamete). The trigger for the transition from asexual to sexual development is unknown, just as the factor responsible for the transformation of invasive merozoites into micro- or macrogametocytes has not been identified (Boyle and Radke 2009).

The onset of gametocyte formation in *T. gondii* appears to be a less controlled process than in other *Coccidia* species. In most *Eimeria*, a fixed number of asexual cycles was identified, followed by the simultaneous entry of most merozoites into the sexual cycle. In contrast, in *T. gondii*, it is not possible to speak of a distinct conversion, as a mixture of asexual and sexual stages is signaled during enteric development. In this species, there are no ultrastructural elements to differentiate merozoites that will transform into sexual stages, nor differences in the host–parasite relationship or between parasitophorous vacuoles (Toulah et al. 2011).

Upon entering the host cell, the merozoite becomes more spherical, losing most of its apical organs (densities and dense granules), retaining the conoid and a few micronemes. This stage (trophozoite) begins to grow, and the mitochondria arranged at its periphery also increase in size. However, at this time, it is impossible to say whether the microorganism will turn into a macrogametocyte or into microgametocytes (Weiss and Kim 2004).

Microgametogony

Currently, there are only a few descriptions of microgametogony (Dubey and Lappin 1998). Initially, differentiation between endogenous proliferative phase and microgametogony is impossible, as both processes involve continuous growth and repeated nuclear divisions. The earliest distinction between the two types of multiplication is based on the differences in the distribution of nuclear chromatin. During schizogony, electron-dense heterochromatin remains dispersed in the nucleus, while in the late stages of microgametogony, it condenses into electron-dense masses at the periphery of the nucleus. In the case of microgametogony, the nuclei move to the periphery of the cell; between them and the plasmalemma, two centrioles and a dense plate are arranged. The centrioles will become the basal bodies of future flagella, which begin to grow by protruding into the parasitophorous vacuole. Although the centrioles differ from those of the metazoans, the flagella have the typical structure with nine pairs of tubules, in the center of which are two microtubules. As the growth of flagella progresses, chromatin continues to condense in the nuclear area closest to the centrioles, with the other portion of the nucleus acquiring a more electron-transparent appearance. In addition, the mitochondria will be arranged in the vicinity of the nucleus. No significant changes in the apicoplast can be identified during the whole process described. The development of the microgamete continues with the protrusion of a portion of the cytoplasm (containing the basal bodies, the electron-dense area of the nucleus and the mitochondria) into the lumen of the parasitophorous vacuole. Along with this process, the nuclear division takes place by separating the electron-dense zone from the electron-transparent one. The electron-dense portion will form part of the microgamete in formation, and the electron-transparent portion will remain in the stem cell as a residual nucleus. Microgametocytes of *T. gondii* produce a relatively small number of microgametes (15–30). Their immature forms remain attached to the stem cell through a narrow cytoplasmic isthmus (Weiss and Kim 2004).

The maturation continues, each microgamete acquiring an elongated appearance and having in its structure an electron-dense nucleus and mitochondria arranged between it and the basal bodies from which the two long flagella start, with posterior orientation. In addition, an electron-dense apical plate and four longitudinally arranged microtubules can be observed. After completing their formation process, the microgametes detach from the microgametocyte, leaving behind a large residual cytoplasmic body (Weiss and Kim 2004).

Macrogametogony

The development of a macrogamete is associated with the growth of the trophozoite and the appearance of a large nucleus, with chromatin dispersed in its mass and with a large nucleolus. The formation of the macrogamete is not accompanied by nuclear division. As the macrogametocyte enlarges, a marked increase in peripherally arranged mitochondria and centrally located apicoplast can be observed. Several Golgi bodies distributed in the cytoplasm can also be visualized. The first distinct organ, specific to macrogametogony, is in the form of a flocculent material, condensed in the dilated portion of the rough endoplasmic reticulum. This material represents the beginning phase of the formation of the wall-forming body type 2 (WFB2—wall-forming body type 2), a formation that received this name due to its role in forming the wall of oocysts. Golgi bodies are often associated with the endoplasmic reticulum membrane that covers WFB2 (Boyle and Radke 2009).

The number of WFB2 increases as maturation continues, while a large number of electron-dense membrane-binding granules derived from vesicles produced by Golgi bodies are also observed. These granules have various sizes and have been designated as type 1 wall-forming bodies (WFB1). The use of immunoelectron microscopy allowed the identification of two populations of electron-dense membrane-binding granules. One of these populations is involved in the formation of the outer shell, which is why it has been designated as a shell-forming body (VFB). Although initially, these granules were named wall-forming bodies of type 1, with the discovery of similar formations in macrogametocytes of the species *Eimeria maxima*, it was considered that the term shell-forming bodies is more appropriate (Weiss and Kim 2004). The type 1 wall-forming bodies are slightly larger than the shell-forming bodies. As the synthesis of the wall and coating bodies occurs, the formation of large numbers of polysaccharide granules and lipid droplets takes place, as well as the increase in size of the apicoplast. The macrogametocyte that has completed its development is considered a macrogamete. This is not a proper criterion of differentiation, but rather one of convenience for distinguishing the stages in development from the mature gamete (Weiss and Kim 2004).

Oocyst Formation

The wall of the *T. gondii* oocyst is a multilayered structure which is extremely resistant to physical and chemical factors. Due to these properties, the wall is essential to the survival of the parasite. Without its wall, the parasite cannot preserve its viability, in the external environment, for long periods.

The wall of the oocyst is a complex structure, consisting of several distinct layers synthesized while the macrogamete is still in the host cell. The first layer is called the outer shell and consists of 2–3 membranes—or layers 1–3 (Weiss and Kim 2004). These result from the release of the contents of the shell-forming bodies, after their fusion with the macrogamete plasmalemma (Ferguson 2009). This occurs during the maturation of the macrogamete and is followed by the simultaneous activation of the secretion of wall-forming bodies of type 1 in the mature macrogamete, with the formation of the outer layer of the oocyst—or layer 4 (Weiss and Kim 2004). This layer initially has a large thickness, and then, via polymerization, acquires the final electron-dense shape and a thickness of 30–70 nm. In the end, the wall-forming bodies of type 2 release their contents which, by fusion, will form the internal electron-transparent layer of the oocyst—or layer 5 (Weiss and Kim 2004). The cytoplasm loses its WFBs during the formation of the oocystic wall and will contain an electron-transparent central nucleus, polysaccharide granules and lipid droplets (Figure 1).

The correct formation and assembly of the oocystic wall is performed only under the conditions of adequate control and sequential secretion of the shell-forming bodies and the wall-forming bodies of type 1 and 2. The data currently available for coccidia are much more accurate. Consider that the outer shell is a specific component of early development and that it is lost with the elimination of oocysts in the feces. Thus, the oocysts are covered by a bilayer wall. The electron-dense outer layer is thinner in *T. gondii* oocysts (Figure 1) than in those of *Eimeria* spp. The formation and polymerization of the inner layer has marked negative effects on the possibility of examining the oocyst ultrastructure. To date, no technique has been developed to examine oocysts of *T. gondii* or other coccidia by electron microscopy. In the last 30 years, regardless of the type of fixation or embedding applied, all such attempts have failed (Weiss and Kim 2004).

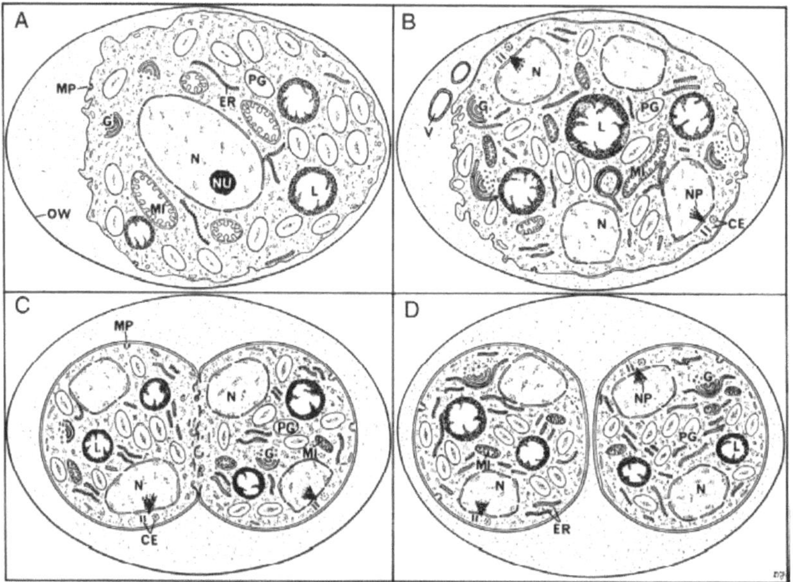

Figure 1. The process of sporoblasts formation. (Ce, centriole; ER, endoplasmic reticulum; G, Golgi body; L, lipids; MI, mitochondrion; MP, micropore; N, nucleus; NP, nuclear pole; Nu, nucleolus; OW, oocyst wall; PG, polysaccharide; V, vacuole). (**A**). Primary sporoblast; (**B**). Formation of four nuclei; (**C**). Cytokinesis; (**D**). Secondary sporoblasts. Source: Reprinted from (Weiss and Kim 2004), used with permission.

The two layers have different roles. The outer layer, which consists mainly of proteins and carbohydrates, guarantees structural strength, and the inner layer, composed mainly of lipids, provides chemical protection by impermeability (even against reagents used in preparations for electro-microscopic examination) (Weiss and Kim 2004).

Fecundation

Sexual development is preceded by the fusion of the microgamete with the macrogamete, resulting in a fertilized zygote. However, this process has not been witnessed thus far. However, mature microgametes and macrogametes can be considered to be released from host cells, with fertilization taking place in the intestinal lumen. This is confirmed by the rare occasions when macrogametes with attached microgametes have been observed. However, the formation of the oocyst wall begins before the macrogamete is released from the host cell. Another peculiarity observed in *T. gondii* is the development of a small number of microgametes, these being relatively few in relation to the number of macrogametes (Weiss and Kim 2004).

Fertilization in *T. gondii* was demonstrated by the identification of cross-fertilized parasites. *T. gondii* is haploid, and the way in which this trait influences the need for fertilization features in a wide range of discussions.

Extracellular Sporulation of the Oocysts

Oocyst is the only stage of *T. gondii* that is able to grow extracellularly—all other processes take place strictly in viable nucleated cells. The oocyst released from the host cell is unsporulated. It consists of a single undifferentiated cytoplasmic mass—the primary sporoblast. In the external environment, asexual development (sporulation) takes place, which results in the formation of two sporocysts. Each sporocyst contains four sporozoites. Initial attempts to observe this process were doomed to failure due to the impossibility of preparing oocysts for ultrastructural examination.

It was then found that examination becomes possible by freezing and cryosection of the oocysts before processing for electron microscopy (Weiss and Kim 2004). The purpose of this method is to section the oocystic wall without destroying the cytoplasmic mass contained in it. The technique proved to be partially inefficient because it resulted in the high destruction of oocysts. However, some of them remained intact and were used to examine the ultrastructural changes that accompany sporulation. Due to these difficulties, the experiments were limited to *Eimeria brunetti*, as a model for the genus *Eimeria*, and *T. gondii*.

The central cytoplasmic mass, called the primary sporoblast, appears to be similar to that of the macrogamete. The cytoplasm contains a single large nucleus and a series of polysaccharide granules and droplets of lipids arranged around mitochondria and the rough endoplasmic reticulum, being circumscribed by a membrane. The *T. gondii* nucleus undergoes two cycles of division, with the formation of four nuclei. Then, the cytoplasm elongates and will be surrounded by two other membranes. This process is followed by the division of the cytoplasmic mass, with the formation in the central area of the invagination of the limiting membranes. Through fusion, these will segment the primary sporoblast into two secondary sporoblasts containing two nuclei each (Figure 2a). Secondary sporoblasts will develop and become more elongated, turning into sporocysts (Weiss and Kim 2004) (Figure 2a). The wall of *T. gondii* sporocysts consists of a material secreted by the cytoplasm, which condenses on one of the limiting membranes. A structure is formed which consists of four plates joined by specialized junctions that cover a thin layer of electron-dense material. This wall is provided with several striations assimilated to repetitive organized protein units that probably provide structural strength and protection in relation to environmental factors (Boyle and Radke 2009). The nuclei can be visualized at both ends of the developing sporocyst, most of the cytoplasm being occupied by polysaccharide granules and lipid droplets. It has been observed that the primary forms of daughter cells form adjacent to the plasmalemma, above each nucleus. The process of daughter-cell formation is similar to that described in endodyogeny, with nuclear division occurring during the posterior growth of the inner membrane complex of each cell. Two daughter cells will form at both ends of the sporocyst (Figure 2b). The inner membrane will develop until the daughter cells are fully formed, enveloping the nucleus, apicoplasts, mitochondria and apical organs (micronemes, rhoptries and dense granules) (Weiss and Kim 2004).

The constitution of daughter cells adjacent to the plasmalemma differs from the process of internal formation associated with endodyogeny and endopolygeny.

In this particular situation, it is observed that an invagination of the plasmalemma takes place with the development of the internal membrane complex in order to form the sporozoite film, similar to the classical schizogony. This will result in the formation of four daughter cells. A small residual cytoplasmic mass will remain in each sporocyst (Figure 2b) (Weiss and Kim 2004).

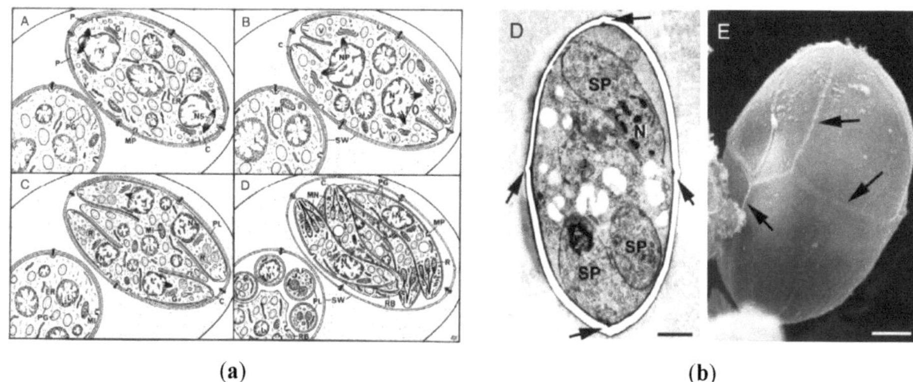

(a)   (b)

Figure 2. (a) A. Developing sporocyst; B. and C. Formation of the anlagen of two daughters at either end of the sporocyst; D. Formation of the sporozoites; (C–conoid; ER–endoplasmic reticulum; G–Golgi body; L–lipids; MI–mitochondrion; MN–microneme; MP–micropore; N–nucleus; NP–nuclear pole; P–the membrane of the future daughter cell; PL–pellicle; PG–polysaccharide; R–rhoptry; RB–cytoplasmic residual body; SW–sporocyst wall; V–vacuole); (b) D. and E. Formation of sporocyst wall resulting in four plates (arrows); N–nucleus formation and the raised junctions. Source: Reprinted from (Weiss and Kim 2004), used with permission.

Excystation

Only a few studies have focused on the ultrastructural characterization of the excystation process (Weiss and Kim 2004). Some experiments resorted to mechanical destruction of the oocyst wall, but there is evidence that excystation also occurs in the absence of mechanical factors. Detection can be stimulated by incubation in a mixture of trypsin and bile salts (sodium bullfighting). This combination causes the tension of the wall, which results in invagination of the edges of the plates at the junction lines (Figure 3a,b). Corresponding to the junctions, a separation of the inner face of the inner layer of the oocystic wall can be observed, which initially remains attached to the outer edges. This connection will be lost, causing the outer membrane of the eye wall to rupture. The plates will separate rapidly and form a roll-like structure, allowing the release of sporozoites (Weiss and Kim 2004) (Figure 3c).

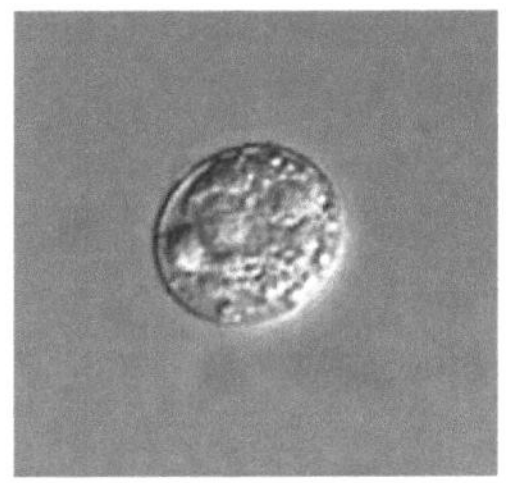 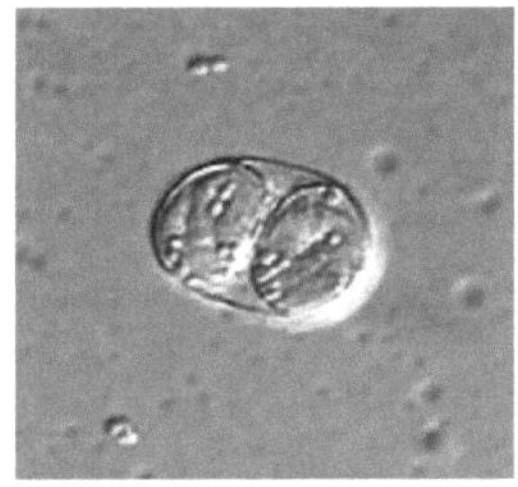 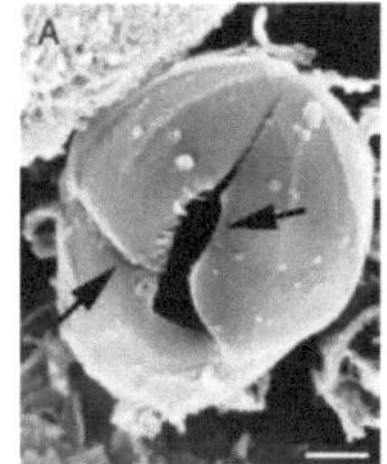 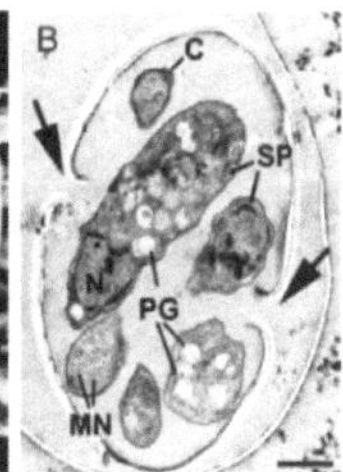

(**a**) Unsporulated oocyst   (**b**) Sporulated oocyst   (**c**) Excystation

Figure 3. *T. gondii* oocysts in an unstained wet mount, viewed with differential interference contrast (DIC) microscopy; (**a**) Unsporulated oocyst of *T. gondii*; (**b**) *Toxoplasma gondii* sporulated oocyst; (**c**) Excystation—separation of the sporocyst wall (arrow). A–B Infolding of the sporocyst wall's plates along the suture lines (arrows); SP–sporozoites; MN–microneme; PG–polysaccharide; C–conoid; N–nucleus. Source: Reprinted from (Weiss and Kim 2004), used with permission.

3.2. Microstructure of the Development Stage of Toxoplasma gondii

3.2.1. Microstructure of Invasive Stages

Tachyzoites Microstructure

In *T. gondii*, four invasive forms can be identified: tachyzoites, bradyzoites, merozoites and sporozoites. Tachyzoites represent the most studied stage of the *T. gondii* cycle because they can be easily obtained in large numbers both in vitro and in vivo (Toulah et al. 2011). This invasive stage is crescent shaped (approximately 2×7 µm), with the anterior extremity slightly sharper (the anterior extremity can be identified depending on the direction of travel) (Figure 4). Tachyzoites consist of a single skeleton (consisting of subpellicular microtubules and conoid), secretory organs (rhoptries, micronemes, dense granules), organs derived by endosymbiosis (mitochondria, apicoplasts), particular organs of eukaryotes (nucleus, endoplasmic reticulum, Golgi apparatus, ribosomes) and specific structures (acidocalcisomes), all circumscribed by a complex membrane structure known as pellicle (Dubremetz and Ferguson 2009).

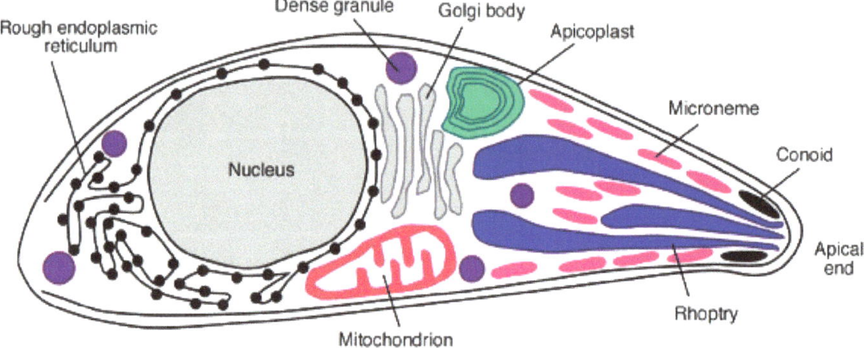

Ultrastructure of a *Toxoplasma gondii* tachyzoite

Figure 4. Cellular elements of tachyzoites (ultrastructure of a *Toxoplasma gondii* tachyzoite). Source: Reprinted from (Ajioka et al. 2001); used with permission. Note: This figure source was updated on 4 June, 2024

The cytoskeleton consists of:

1. Two apical rings located below the plasma membrane of the apical extremity of the parasite. These two rings are formed of a narrow band of electron-dense material.
2. The conoid has the appearance of a truncated and hollow cone formed of spirally wound tubulin fibers.
3. Two polar support rings surrounding the tip of the conoid. The outer ring is made of dense material. The lower ring is made of a material that anchors the subpellicular microtubules.
4. A pair of adjacent intraconoidal microtubules extending over an extremely short length (less than 1 µm) into the apical cytoplasm (Weiss and Kim 2004).

They have a distinct membrane complex that envelops the infectious stages (Craver et al. 2010). It consists of a superficial (outer) membrane unit—the plasmalemma that completely covers the microorganism—and an inner layer of two tightly bonded membrane units arranged at a fixed distance (approximately 15 nm) from the plasmalemma. The internal membrane complex (IMC) consists of fused plates that are formed by flat vesicles, derived from the endoplasmic reticulum–Golgi apparatus system. The inner layer is interrupted by circular pores at the anterior end (outer polar ring), where the conoid protrudes, and at the posterior extremity. Both sides of the inner membrane complex have their protoplasmic faces covered by intramembrane particles (IMPs); 22 lines with high density, corresponding to the subcellular microtubules. IMPs are arranged at a distance of about 30 nm. The organization of IMP in the apical plate is extremely different from that observed for the other plates, which suggests that the apical region is characterized by a distinct molecular structure (Ferreira-da-Silva et al. 2009).

The pellicle has micropores located in the apical half of the cell, usually anterior to the nucleus. A micropore consists of a circular invagination of the plasmalemma (approximately 150 nm in diameter), which passes through the inner membrane complex. The latter folds and forms an electron-dense sleeve around the invagination. It is believed that these structures have a role in the uptake of substances (Weiss and Kim 2004).

In *T. gondii*, three distinct types of secretory organs have been identified (which secrete specific proteins), the number and shape of which vary according to the invasive stage (Weiss and Kim 2004). The first type—micronemes—is in the form of small sticks (250 × 50 nm) and is located mainly in the apical region, behind the conoid. Micronemes are characterized by homogeneous electron-density. The second category is represented by the rhoptries, which are organized as a group of elongated, clover-shaped organs, and which are arranged between the conoid and the nucleus. The rhoptries have a narrow, long neck (up to 2.5 μm long) and a bag-like body, measuring 0.25 × 1 μm, located in the posterior area. Their content is electron dense, except for the widened portion which, depending on the stage, may be electron dense or labyrinthine. The third type of organelle, dense granules, is found throughout the parasite's cell and in greater numbers in the posterior region. Dense granules are spherical structures with a diameter of 0.3 μm and electron-dense content (Dubremetz and Ferguson 2009).

The nucleus occupies the central or basal area of the cell, depending on the invasive stage. It is usually flattened in the upper area, where the Golgi device is located. It contains a central nucleolus and small agglomerations of electron-dense heterochromatin scattered throughout the nucleoplasm. The nuclear membrane is provided with a large number of nuclear pores, and its outer face is covered with ribosomes (except for the upper part, where the Golgi apparatus is located). The nuclear membrane is found in the continuity of the rough endoplasmic reticulum, which extends into the cytoplasm of tachyzoites (Boyle and Radke 2009).

A layer of clear vesicles 70 μm in diameter can be seen over the upper part of the nucleus, some appearing to sprout from the nuclear membrane. This layer is covered by three or four flat Golgi cisterns which, in turn, have the upper area covered with vesicles with various contents and sizes.

The use of a specific processing technique has made it possible to identify, in the posterior portion of the tachyzoites or in the vicinity of the nucleus, one or two vesicles of about 200 nm, containing one or more droplets or crystals of different sizes. These vesicles have been called acidocalcisomes, and their black content is considered to be calcium bound to pyrophosphate or polyphosphates (Weiss and Kim 2004).

The profile of several mitochondria with a width of 0.5 μm and varying length can be observed in different locations, usually above and below the nucleus. This profile represents a section through an unbranched and elongated mitochondrion. Mitochondria are characterized by a typical structure for the *Apicomplexa* branch, with crystals in the form of bulbs.

Above the Golgi apparatus are the apicoplasts. These organelles, limited by multiple membranes, have been identified since the 1960s, but only recently have they been shown to be typical plastids (Weiss and Kim 2004). Because they appear to be a specific feature of the *Apicomplexa* phylum, with the exception of *Cryptosporidium* spp., the term "apicoplasts" has been proposed and adopted. These organelles have a relatively constant shape, which is almost unaffected by the infectious stage; a diameter of about 500 nm, covered by four membranes; and contain a granular and filamentous material in which ribosomes can be observed. The origin of the four membranes that surround the apicoplasts is still under discussion.

Difference between the Invasive Stages

The infectious stages represented by tachyzoites, bradyzoites, merozoites and sporozoites show differences in the number of apical organs, the shape and electron-density of the rhoptries, the intracellular location of the nucleus and the presence or absence of polysaccharide granules. In tachyzoites and merozoites, the nucleus has a central position, and in bradyzoites (Figure 5) and sporozoites it is placed basally (Ferreira-da-Silva et al. 2009).

An additional structure is the polysaccharide granule. Polysaccharide granules are oval structures (250–180 nm), with variable electron density, located in the apical and basal cytoplasm of the parasite. They contain an unusual form of carbohydrates that, biochemically, are more similar to plant amylopectin than animal glycogen (Weiss and Kim 2004). These granules are present in small numbers in tachyzoites and merozoites and very well represented in bradyzoites and sporozoites. It is considered that polysaccharide granules would have the role of storage and energy source necessary for bradyzoites and sporozoites during the passage between hosts (Dubremetz and Ferguson 2009). The most marked difference between the infectious stages refers to the apical organs, these differences being marked in Table 3 (Weiss and Kim 2004).

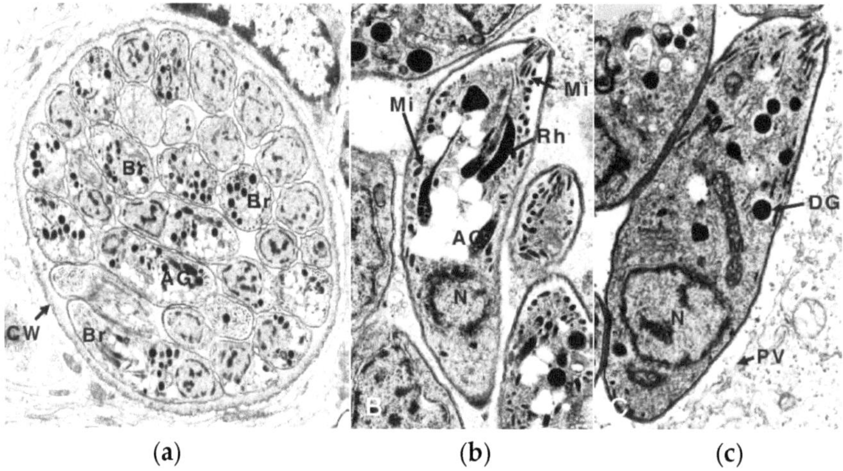

(a)   (b)   (c)

Figure 5. Parasitic stages of *Toxoplasma gondii*: (a) Tissue cyst: CW, cyst wall; Br, bradyzoites; AG, amylopectin; (b) Bradyzoite structure: Rh, rhoptry; MI, micronemes; N, nucleus; AG, amylopectin; (c) Tachyzoite structure: few amylopectin; DG, dense granules; M, mitochondrion; N, nucleus; **PV**, parasitophorous vacuole. Source: Reprinted from (Weiss and Kim 2004), used with permission.

Table 3. Morphological differences between stages of Toxoplasma gondii.

The Parasitic Stage	Nucleus	Micronemes	No. of Rhoptries	Aspect	Dense Granules	Polysaccarides Granules
Tachyzoites	Central	Few	5–12	Labyrithine	Numerouse	Few
Bradyzoites	Basal	Numerouse	5–10	Compact	Numerouse	Numerouse
Merozoytes	Central	Few	3–5	Compact	Few	Absent
Sporozoytes	Basal	Numerouse	5–10	Labyrithine	Numerouse	Numerouse

Source: Authors' compilation based on data from (Weiss and Kim 2004), used with permission.

3.2.2. Morphology and Biology of Tissue Cysts and Bradyzoites

Tachyzoites (tachos = fast) denote the rapidly growing stage of *T. gondii*, also known as endozoytes or trophozoites. Bradyzoites (brady = slow, slow), also called cystozoites, represent the stage of the parasite located in tissue cysts and are considered to be a slow-replicating form. Both forms replicate inside the parasitophore vacuoles included in the host cells, structures that then differentiate, depending on the stage of development, into specific vacuoles for tachyzoites (pseudocysts) or for bradyzoites (tissue cysts) (Craver et al. 2010). Tissue cysts are intracellular structures within which bradyzoites divide by endodiogeny (a replicative mechanism identical to that identified for tachyzoites) (Boyle and Radke 2009).

Bradyzoids in mature cysts can enter into the atency stage of the cell cycle, becoming essentially differentiated, non-replicating organisms with DNA content (Zintl et al. 2009). Meanwhile, in mature cysts, occasionally degenerated bradyzoites can be observed. In mature bradyzoites, the absence of segregation and loss of apicoplasts has also been reported in vitro (Dubremetz and Ferguson 2009).

The size of the cysts varies depending on their age, the type of host cell, the *T. gondii* line and the cytological method applied for measurement (Craver et al. 2010). Thus, for mature cysts located in the brain, a spherical shape and a diameter of 50–70 µm were reported. They contain about 1000 crescent-shaped bradyzoites with dimensions of 7×1.5 µm (Zintl et al. 2009). Cysts located in the muscles have an elongated shape, reaching 100 µm in length (Weiss and Kim 2004).

Immature and old cysts can be easily identified based on ultrastructural characteristics (Craver et al. 2010). The structure of tissue cysts remains relatively unchanged in the observation period from 3 months to 24 months after infection (approximately the average lifespan of host mice). The first important observation is that during this period, the cyst is maintained in a viable host cell (Zintl et al. 2009). Mature cystic formations were initially thought to be extracellular, but ultrastructural examination, performed on fine sections through the cytoplasm of host cells, demonstrated the opposite (Weiss and Kim 2004). Masking of tissue cysts by host cells could be an explanation for the lack of immune response to them. Although cysts can develop in any parenchymal organ (e.g., lung, liver, kidney), it has been observed that they are located primarily in neural tissue (e.g., in the brain and/or eye nerves) or muscle (e.g., in skeletal and/or cardiac muscle). Due to the fact that the host cells retain only a small amount of cytoplasm, it becomes difficult to specify their belonging to a certain cell type (Boyle and Radke 2009).

The wall of the tissue cyst or bradyzoite's parasitophorous vacuole is elastic, thin (<0.5 μm thick), weak PAS-positive and silvery (this feature depends on the staining method applied). On phase contrast microscopy, the wall appears phase bright. Vacuoles frequently contain an odd number of clover-shaped organisms (asynchronous division) (Craver et al. 2010).

Bradyzoites differ ultrastructural from tachyzoites in that they have a posteriorly located nucleus and possess solid, often loop-shaped, numerous micronemes and polysaccharide (amylopectin) granules (Zintl et al. 2009). Lipid bodies are absent in bradyzoites but are numerous in sporozoites and occasionally observed in tachyzoites. The content of mature bradyzoites is electron dense, in contrast with the labyrinthine ruptures reported in tachyzoites and immature bradyzoites. Bradyzoites are PAS positive (they turn red), and tachyzoites are PAS negative (Ferreira-da-Silva et al. 2009).

Bradyzoites are generally thinner than tachyzoites and more resistant than these to treatment with pepsin and acid (they survive 1–2 h in the mixture of pepsin with HCl, compared to 10 min as tachyzoites resist) (Dubremetz and Ferguson 2009).

The formation of the wall and the matrix of tissue cysts represent early elements of the differentiation process in bradyzoites (Craver et al. 2010). One of the most important functions of the cystic wall and matrix is to protect bradyzoites from inadequate environmental conditions, such as dehydration. In addition, these structures represent a real physical barrier for the host's immune system, much of this function being provided by carbohydrates in the parietal structure (Weiss and Kim 2004).

The cyst wall is actually a change in the membrane of the parasitophorous vacuole formed by the parasite inside the host cell—in other words, the tissue cysts are located intracellularly. The membrane of the parasitophorous vacuole has a wrinkled appearance on electron microscopy and appears to be associated with the precipitated underlying material, thus forming the cystic wall (Boyle and Radke 2009).

In Vitro Development of Bradyzoites and Tissue Cysts

The in vitro development of tissue cysts was reported more than 40 years ago (Weiss and Kim 2004), but due to the similar morphology of bradyzoites and tachyzoites on classical microscopy, the study of the differentiation process was very difficult until specific antibodies—anti-bradyzoites—were obtained. It has been observed that the formation of tissue cysts is characterized by a marked tissue tropism (Zintl ct al. 2009). The two types of tissue in which the vast majority of tissue cysts have been observed are striated muscles, but also in the heart and the central nervous system. Variations can also be observed depending on the host species (Craver et al. 2010). In pigs, most tissue cysts are located in the muscles, and in mice, they are mostly located in the brain. This finding contradicts the observations made in vitro culture, a situation in which almost any cell type can be a host cell during endodiogenesis (Zintl et al. 2009).

Stage conversion was examined in detail, with mice as animal models (Ferreira-da-Silva et al. 2009). At 12–15 days after the oral infection, the brain lesions were visualized, indicating the presence of parasites that turned into tachyzoites and those that formed early tissue cysts (Figure 6a,b). It has been found that only a small number of tachyzoites perform the conversion (Boyle and Radke 2009).

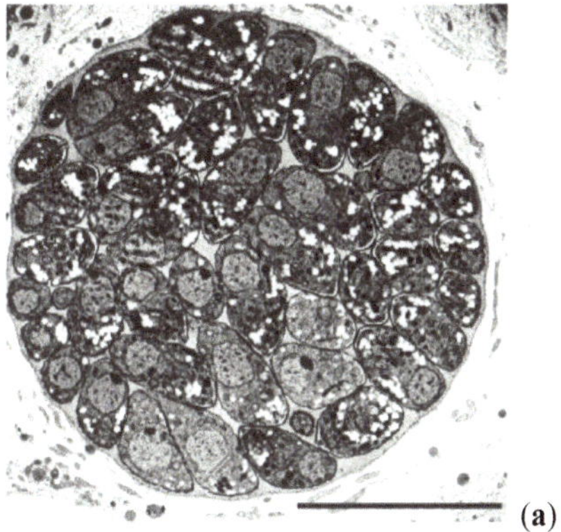
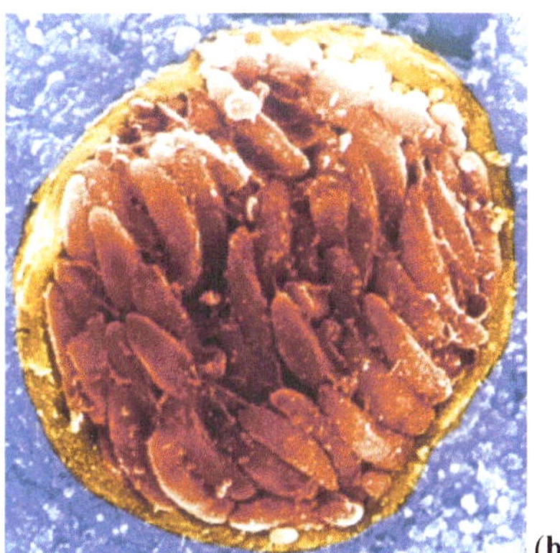

Figure 6. (a) In vitro developed tissue cyst. Source: Reprinted from (Weiss and Kim 2004), used with permission; (b) The protozoan *Toxoplasma gondii*, tissue cyst in brain. Source: Authors' compilation based on data from D. Ferguson, Oxford University. Available online: https://www.uchicagomedicine.org/forefront/biological-sciences-articles/what-does-it-mean-when-2-billion-people-share-their-brain-with-a-parasite (accessed on 27 November 2019).

However, the contribution of the host cell to the conversion of parasitic stages remained unexplained. Two teams of researchers (Weiss and Kim 2004) reported that exposure of extracellular tachyzoites to stress conditions (pH = 8.1) causes differentiation in bradyzoites. Changes in cultivation leading to bradyzoite formation relate to temperature (43 °C), pH (pH 6.6–6.8 or 8.0–8.2), exposure to various substances (sodium arsenite) and deprivation of certain nutrients (arginine). Changing the pH is one of the most common methods of inducing differentiation in bradyzoites in the laboratory. However, most differentiation-inducing agents exert their effects, to a greater or lesser extent, on the host cell, which is why it is very possible that the alterations induced in the host cell will have a considerable impact on the parasite. So, this may be an indirect mechanism of differentiation (Zintl et al. 2009).

The interconversion of these two stages—bradyzoite ⇔ tachyzoite—is a fast and obligatory process for differentiation, which occurs at the moment or at a short interval after the invasion and the formation of the parasitophorous vacuole (Ferreira-da-Silva et al. 2009). Tissue cysts continue to grow over the next three weeks, with bradyzoites multiplying by endodiogenesis. Initially, in the process of proliferation, a large number of parasitic elements are involved, and then the proportion of bradyzoites in division decreases in the first 28 days post-infection, and in the next three months, only very few are needed to divide. The tissue cyst enlarges in the early stages of development, invaginations deepen and the layer of amorphous material becomes clearer (Craver et al. 2010). Bradyzoites located in early tissue cysts have ultrastructural characteristics similar to tachyzoites, especially in terms of the labyrinthine appearance of rhoptries. This proves that, despite the typical structure of tissue cyst and the expression of specific markers of bradyzoites, the change in ultrastructural characters occurs later. It usually takes 21–28 days for typical bradyzoites to be identified. In addition, many parasitic elements in division have a more electron-transparent cytoplasm with few organs (Dubremetz and Ferguson 2009).

Cysts Breaking in Immunocompetent Hosts

One of the clinical problems in immunocompromised hosts is the resurgence of infection with reverse conversion, which results in the formation of tachyzoites with the ability to multiply rapidly, accompanied by massive tissue destruction. To assess the status of immunocompetent hosts, the brain was examined in immunocompetent mice with chronic infection. It has been observed that a small percentage of tissue cysts are ruptured, regardless of the period elapsed from the time of infection. The first change is the death of the host cell. Exposure of the parasitic antigen in the cystic wall triggers a massive and rapid cellular immune response (accumulation of monocytes and even neutrophils) (Boyle and Radke 2009).

Effector cells can be seen around the seemingly intact tissue cyst. The rupture of the cystic wall will allow the invasion of macrophages inside it. Macrophages will phagocytose bradyzoites, a process that involves their fusion with lysosomes (formation of phagolysosomes), with the destruction of parasitic elements. The described immune response results in small inflammatory lesions (microglial nodules), with some apoptosis of local cells, but with a low "price" for the host. Bradyzoites are destroyed before replicating or turning into tachyzoites (Ferreira-da-Silva et al. 2009). These findings were confirmed by immunocytochemical analysis of lesions in immunocompetent individuals, processes that appear to be lacking in immunosuppressed individuals, thus allowing infinite multiplication of tachyzoites and destruction of as many host cells (Weiss and Kim 2004).

3.3. Cultivation of Toxoplasma gondii

It has been often claimed that *T. gondii* is an easy-to-cultivate species, which is why it has become highly valued as a model for molecular study. However, it should be noted that normally, only tachyzoites and bradyzoites can grow in cell cultures (Zintl et al. 2009). More recently, in vitro activation of tissue cyst formation has been successful, but all attempts to reproduce the development of coccidian stages in laboratory conditions have failed. Oocysts can be obtained by feeding free cats with cysts from infected mice (Dărăbuş et al. 2006).

T. gondii can be cultured in laboratory animals, chicken embryos and cell cultures (Zintl et al. 2009). Mice, hamsters, guinea pigs and rabbits are susceptible species. Mice are more commonly used because they are more receptive and do not become naturally infected when are raised in the laboratory, with commercially dried food, uncontaminated with cat feces (Cosoroabă 2005).

The *T. gondii* protozoan can be grown and maintained in tissue cultures by passing twice a week on Vero cells (African green monkey kidney cells). According to other authors (Nerad and Daggett 1992), preputial fibroblasts are one of the cell lines that can serve as a host for *Toxoplasma gondii*. The cell line can be maintained in the culture medium composed of L-amino acids, vitamins and Earle salt (Nerad and Daggett 1992). The culture medium can be supplemented with penicillin, strep-tomycin and fetal bovine serum. *Toxoplasma gondii* cannot be cultured on acellular media.

Toxoplasma gondii, maintained both in vitro and in vivo, can be cryopreserved in liquid nitrogen for long-term preservation. Because the parasite is pathogenic to humans, the main issues to be considered are those related to operator safety. Tachyzoites can be obtained in mice or cell cultures according to the method described above (Daggett and Nerad 1992).

3.3.1. In Vitro Development of Tachyzoites

Regardless of the type of host cell, tachyzoites develop in vitro, following the principles described above (Section 3.1.1.2—Tachyzoites biogenesis).

Tachyzoites of certain strains of *T. gondii* grow in the peritoneal cavity of mice, sometimes producing ascites, and develop in most other tissues after intraperitoneal inoculation with any of the three infectious stages of *T. gondii*. Virulent strains usually cause mice to become sick and sometimes die in 1–2 weeks. In experimental infections, cysts obtained from mice infected for at least three months are used (Zintl et al. 2009).

Toxoplasma gondii tachyzoites can be multiplied on many cell lines. Although tissue cysts can develop in cell cultures of most strains of *T. gondii*, their number is smaller than in mouse infections (Weiss and Kim 2004).

3.3.2. In Vitro Development of Bradyzoites

The in vitro development of bradyzoites was first reported in the 1980s in astro-cytes (Weiss and Kim 2004), and the technique of inducing conversion at this stage was discovered in the 1990s. Unlike the situation reported in vivo, characterized by an obvious selection of the host cell type (neurons, muscle cells), virtually any cell type can play a host role for the development of tissue cysts in vitro. In some studies, the process of tissue cyst formation has been described according to the observations of in vivo evaluations, namely cystic formation from the moment the parasite enters the host cell (Craver et al. 2010). Other studies (Weiss and Kim 2004) suggest that, before the vacuole acquires the characteristics of a tissue cyst, a development similar to that observed in tachyzoites occurs.

Tissue cysts are obtained by injecting tachyzoites, bradyzoites and oocysts into mice. To obtain these cysts from mice inoculated with virulent strains, it is necessary to administer an anti- *T. gondii* chemotherapeutic product to prevent death due to acute toxoplasmosis, before the formation of tissue cysts. For the treatment of acute toxoplasmosis in mice, sulfadiazine is used successfully. Cysts are visible in the brains of mice 8 weeks after infection (Zintl et al. 2009).

Differentiation in the form of bradyzoite appears to be associated with slowing of the cell cycle. It is very likely that a series of signals will become suitable elements for inducing bradyzoite formation. The development of bradyzoites appears to be a differentiation response mediated by stress factors, leading to metabolic adaptation. What is certain is that, during differentiation, the whole set of genes specific to bradyzoites takes place. The products of these genes include enzymes with a role in metabolism, surface antigens, secretory antigens (including rupture proteins) and cystic wall components (Dubremetz and Ferguson 2009).

Investigations into the biology of bradyzoites and the differentiation of tachyzoites into bradyzoites have been accelerated both due to the development of in vitro techniques for the production of bradyzoites and against the background of genetic tools available for the manipulation of *T. gondii* (Zintl et al. 2009).

Studies on the mechanism by which the development of bradyzoites is triggered may serve to identify new therapeutic strategies for the control of toxoplasmosis. These benefits may ultimately result in the complete cure of the infection by eliminating the latent cystic stages. In addition, genetic strategies that prevent the formation of tissue cysts may be useful for developing a vaccine line (Weiss and Kim 2004).

3.4. Molecular Diversity of Toxoplasma gondii

The first genotyping studies of *T. gondii*, published 15–20 years ago, concerned the description of a population structure of *T. gondii* divided on three main lines, designated according to virulence in relation to the murine animal model as type I, II and III (Frazão-Teixeira et al. 2011). The interpretation at that time of population structure and universal genetic diversity was distorted by the fact that it relied exclusively on the analysis of a limited number of strains and laboratory isolates, obtained mainly from humans and domestic animals in France and the United States. Consequently, several isolates proved to be different from these three main types and were found to usually originate in other geographical areas (Beck et al. 2010). Subsequent studies performed on several isolates from different countries and multilocular genotyping have shown the existence of a much more complex population structure of *T. gondii*, which encompasses greater genetic diversity than was initially considered. It was observed that genotypes that could not be included in the three main types came mostly from other continents. Consistent with the data obtained, the view on the simplistic population structure of *T. gondii* was revised. All these advances in understanding the genetic diversity of *T. gondii* have been made possible by the development of genetic markers (Frazão-Teixeira et al. 2011).

Multilocular typing methods, based entirely on the SAG2 locus and which did not allow the identification of atypical strains or those resulting from genetic recombination, have become completely useless. Multilocus microsatellite analysis has been appropriately correlated with genotypes identified by single-gene PCR-RFLP. Microsatellites are mainly useful for the analysis of population structure or for the individual identification of isolates (Beck et al. 2010).

3.4.1. Population Structure on Toxoplasma gondii: The Three Main Lines

Sequencing of numerous individual genetic regions confirmed the clonal nature of the three main lines of *T. gondii* that were originally described (Table 4).

The rate of polymorphism over the entire genome between the three main lines was estimated to be approximately 0.65% (Frazão-Teixeira et al. 2011). These lines are characterized into type I, II and III SNP (single nucleotide polymorphism). Large chromosomal regions are dominated by one of three types of SNPs. The asymmetric distribution of types I, II and III of SNP shows that types I and III represent the second, respectively, the first generation resulting from the crossing of the type II line with one or two ancestral lines (da Silva et al. 2011).

The remarkable absence of diversity within each of these lines and the small divergence between them are clear evidence that the three clonal archetypes have emerged relatively recently as dominant lines. From the calculation of the rates of neutral mutations, it was found that they were detached from a common ancestor 10,000 years ago (Beck et al. 2010). The fact that microsatellite loci are also characterized by a low polymorphism between the three lines also supports the idea of recent detachment, at least on the continents where they predominate.

One of the biological explanations for this clonal population structure is given by the infectivity of tissue cysts, which allows oral transmission to other intermediate hosts without the sexual cycle. It has been hypothesized that this property, unique among prey–predator cycle coccidia, occurred simultaneously with the recombination that generated the three predominant clonal lines (Table 4) (Beck et al. 2010).

Table 4. Biological and epidemiological characteristics of *Toxoplasma gondii* genotypes.

Genotypes of *Toxoplasma gondii*	Characteristics
Type I	-Solarely isolated. -Very virulent for mice (but not for rats): causes death in mice inoculated with less than 10 tachyzoites. -In cell cultures: high growth rate, but a low conversion rate of tachyzoites to bradyzoites. -Increased possibility of long-distance migration; crosses polarized membranes or extracellular matrix. -Increased penetration rate in own or submucosal lamina "ex vivo".
Type II	-The most frequently isolated type in Europe and North America. -Nervirulent for mice: causes chronic infection with persistent tissue cysts. -In cell cultures: low growth rate, but a high conversion rate of tachyzoites to bradyzoites.
Type III	-Less common than type II in Europe and North America. -Nevirulent for mice, although death in mice, accompanied by neurological signs, may occur a few weeks or months after inoculation.
Recombinant, atypical or exotic types	-Most frequently isolated outside Europe and North America (in South America, Africa, Asia). -Usually more virulent for mice than type II (death occurs 2–3 weeks after inoculation), but variations in virulence and other biological properties reflect the difference in inherited genes.

Source: Authors' compilation based on data from (by Maubon et al. 2008), used with permission.

3.4.2. Strains Not Belonging to the Three Major Lines

Since the application of multilocular genotyping over a wider geographical area and to a larger number of host types, strains that differ from the three major lines have been described more and more frequently (Beck et al. 2010). The multilocular and multichromosomal genotyping of the strains mentioned by PCR-RFLP or microsatellite markers showed that the vast majority of them have alleles of type I, II and III, identical to those identified in the major lines, but with different segregation in the analyzed loci. All these strains are considered the result of recombination of major lines (da Silva et al. 2011). These genotypes have been identified less frequently, suggesting a lower proliferation capacity, but the extension of the analysis outside Europe and North America may change this view. In fact, predominant haplogroups have already been described in these atypical genotypes, which may be associated with new clonal lines (Beck et al. 2010). From the above, it can be concluded that the population structure of *T. gondii* is governed by recombination that generates the rapid expansion of new clonal types. For example, the multilocular genotyping of 125 isolates from different regions of Brazil allowed the identification of 48 genotypes, and the phylogenetic network between them indicates a high recombination rate (Beck et al. 2010). Four of these genotypes have been detected in several areas (arranged at distances of 3000 km) and can be considered common clonal lines that have resulted from genetic exchanges and have spread successfully in Brazil.

Combining, in phylogenetic analysis, the strains from Europe, North America and South America, Khan et al. 2007 (cited by Beck et al. 2010) differentiated 11 haplogroups (haplotypes) which also include the three main lines (renamed haplotypes 1, 2 and 3) and other clonal groups in South America (haplotypes 8 and 9) (Beck et al. 2010). These haplotypes are characterized by marked geographical segregation between South and North America. Some of these genotypes possess unique polymorphism for loci, a feature that has not been identified in the three main lines (da Silva et al. 2011). Phylogenetic analyses have ruled out the possibility of kinship with major lines. To emphasize the differences from the main lines, these strains have been called "atypical" (Beck et al. 2010). However, the cataloging of an isolate may be partially real because it depends on the number of markers used in genotyping and their ability to distinguish. All these names and classifications are provisional, expecting a more adequate definition of the population structure in *Toxoplasma*. The most divergent strains were isolated from severe cases of human toxoplasmosis reported in the Amazon rainforest of French Guiana—unique multiple polymorphisms have been reported, and microsatellite multilocular sequencing has shown that each strain has a unique multilocular genotype.

The naturally occurring strains are not only extremely divergent, but are also characterized by great genetic diversity. Frequent genetic changes, with the formation of a wide range of recombination, seem to be more characteristic of *T. gondii* isolates in areas where animal husbandry is recent and extensive, with recent domestication of cats (Frazão-Teixeira et al. 2011). Examples include *T. gondii* isolates from the Amazon and those from wildlife (bears, deer) in Canada. In Europe and other parts of North America (other than Canada), the development of the domestic and peridomestic cycle (between cats, rodents, birds and farm animals) has accentuated the possibility of clonal expansion and had negative effects on genetic diversity. The estimated time for the expansion of the three main clonal types coincides with the beginning of cat farming and domestication in Southwest Asia and the Middle East (Beck et al. 2010).

3.4.3. Geographical Distribution of *Toxoplasma gondii* Genotypes

Over 95% of isolates originating in North America and Europe belong to the three main lines, but their distribution is not uniform (Frazão-Teixeira et al. 2011). In France, more than 600 isolates belonging almost exclusively to type II have been sequenced. This type is responsible for more than 90% of cases of congenital toxoplasmosis in humans and corresponds to all isolates identified in a wide range of animals. Type II is also predominant in Germany and Poland. The predominance of type II has been confirmed by extensive seroepidemiologic studies in several other European countries. In Spain, type I, determined by monolocular typing, was described more frequently in cats (26%). Preliminary studies show that type III may be more common in Portugal. Although the number of isolates analyzed by multilocular genotyping is still low, the population structure of *T. gondii* seems to be homogeneous in Europe, the non-clonal forms being the exception (Beck et al. 2010).

The preponderance of main lines (especially type II) has also been reported for North America. Recently, a wide variety of non-clonal isolates have been discovered in wild and domestic animals. The more recent introduction of cat farming and domestication on the North American continent could be an explanation for the greater diversity of strains compared to Europe. Clonal types I, II and III are rare in Central and South America, being replaced by various haplotypes, some of which are considered clonal. The increased diversity observed in South America is even more noticeable in the Amazon. The divergence between South and North America is a controversial topic. Lehmann et al. (2006) (cited by Beck et al. 2010) hypothesized that South American strains are older than North American ones. Contrary to this view, Sibley and Ajioka (2008) (cited by Beck et al. 2010) state that *T. gondii* arrived in South America with cats that migrated from the northern continent and became geographically isolated, suffering drift from subsequent genetics (Beck et al. 2010). Early reports of Asia show more limited genetic diversity, with some haplotypes that are also common in South America (da Silva et al. 2011). In Africa, two of the main lines (type II and III) have been identified mainly in Egypt and Uganda, suggesting their possible predominance on the continent. Studies based on multilocular microsatellite analysis of other strains isolated from African patients (mainly from Central and West Africa) show the presence of a fixed combination of alleles belonging to types I and III, and the hypothesis of an African clonal type probably related to the type of BrI detected by Pena et al. (2008) (cited by Beck et al. 2010) in Brazil (Beck et al. 2010).

Over the years, real progress has been made in the field of population genetics and biodiversity in *Toxoplasma*. However, large-scale molecular and epidemiological studies are needed to elucidate the global spread of different genotypes or the influence of various early (continental drift, feline migration) or more recent (domestication of cats and farm animals, genetic exchanges and migration with cats and pests) on the evolution of *T. gondii* (da Silva et al. 2011). All these aspects go far beyond the theoretical interest, because genetic recombination occurring in wildlife hosts may result in the emergence of strains equipped with pathogenic mechanisms that facilitate the rapid expansion and installation of the disease (Frazão-Teixeira et al. 2011).

4. Toxoplasmosis Epidemiology

4.1. The Receptivity in Toxoplasmosis

The spread in the world is uneven. Some of the causes could be the structure of the fauna, the environmental conditions and the degree of culture. The incidence seems to be higher in hot and humid areas than in dry and cold ones. In some intermediate hosts, a certain seasonality of evolution could be observed. Thus, in herbivores, the prevalence is higher in spring, while in pigs, it is higher in autumn and winter (Venturini et al. 2004; Wolf et al. 2005; Dărăbuş et al. 2006; Van der Giessen et al. 2007; Yu et al. 2007).

Among the definitive hosts, the most important are the domestic and wild cat, the ocelot, the Bengal tiger, the lynx and the margays. The susceptibility of intermediate hosts to infection is very high for mouse, guinea pig, golden hamster, rabbit; great for sheep, dog, fox, man; average in pigs; and lower in cattle and horses. The rat appears refractory to infection (Azevedo et al. 2005; Djurković-Djaković et al. 2005; Dubey 2006; Figueroa-Castillo et al. 2006; Naves et al. 2005). Some mouse lines are more receptive than others, and the severity of infection in individuals in the same line may be different. Some species are genetically resistant to clinical toxoplasmosis. For example, adult rats do not get sick, while young rats can die from toxoplasmosis. Mice of any age are susceptible to clinical infection with *T. gondii*. Adult dogs, like adult rats, are resilient, while puppies are very sensitive. Cattle and horses are among the species most resistant to acute toxoplasmosis, while some marsupials and monkeys in the New World are most susceptible to infection. Nothing is known about the genetic determinism of susceptibility in clinical toxoplasmosis of higher mammals, including humans (Furuta et al. 2001; Garcia et al. 2005; Salant et al. 2009a, 2009b).

4.2. Sources of Infection in Toxoplasmosis

The main source of infection for intermediate hosts is the cat, which eliminates oocysts by contaminating the environment. A cat can eliminate about 20 million oocysts daily. The habit of cats to bury their feces in grain depots creates the possibility of transmitting oocysts to farm animals. *T. gondii* oocysts can be dispersed in the environment by atmospheric factors wind, rain or surface water (Dubey 2004). Hay, straw or grass that has been contaminated with cat feces is considered a source of infection for animals. In addition, oocysts can be spread by earthworms, coprophagous mites or manure used as fertilizer. In the intestines of kitchen beetles, which can act as vectors, oocysts remain infectious after 19 days. Other cats may also be a source of infection, but they eliminate fewer oocysts compared to the domestic cat (Chinchilla et al. 1994; Dubey 2002; Dubey et al. 2005c; Mucker et al. 2006).

People can become infected through contact with contaminated soil, for example through gardening, and oocysts are isolated from soil samples from different parts of the world. It is also assumed that dog fur, which comes in contact with cat feces, may play a role in transmitting oocysts to humans (Ryser-Degiorgis et al. 2006).

Organs and tissues, derived from the approximately 350 species of receptive vertebrates, which are infected with tachyzoites and bradyzoites are sources of infection for the final hosts—cats. They can also be sources of infection for intermediate hosts (Belbacha et al. 2004; Figliuolo et al. 2004; Silva et al. 2007). Other sources of infection are wild rodents, small birds and wild rabbits, which are believed to be the natural source of infection for cats (Hill et al. 1998; Dubey 2002; Figueroa-Castillo et al. 2006).

Cases of toxoplasmosis in humans and animals caused by infestation with contaminated water or soil are increasingly cited in the literature (Isaac-Renton et al. 1998; Slifko et al. 2000).

Dumètre et al. (2008), in France, tested two methods for inactivating the oocysts from drinking water. They underwent UV water testing, as well as ozone treatment. It has been observed that *T. gondii* oocysts in water have been inactivated after UV treatment, but not after ozone treatment (Dumètre et al. 2008).

Knowing that ozone treatment and water chlorination, in addition to being ineffective in inoculating oocysts, are also harmful to the released products, it was concluded that UV treatment is the most effective and least harmful method of inactivating *T. gondii* in weeds. In endemic areas (such as France), this method could be used in the prophylaxis of toxoplasmosis in humans and animals (Dubey 2004).

Afonso et al. (2008) intended to identify the defecation sites of cats in Lyon, France. They gave the cats food marked with dyes, and their feces was collected. In total, 72 fecal samples were recovered from 24 cats. Along with the feces, soil samples were collected and examined by PCR to determine oocysts. In 3 out of 55 soil samples, and in the second experiment, in 8 out of 62 soil samples, the presence of *T. gondii* oocysts could be identified. These data confirm the risk of contamination of humans and animals with environmental oocysts (Afonso et al. 2008).

Dumètre and Dardé (2005) state that immunomagnetic separation would be an effective method for determining *T. gondii* oocysts in water, but until then, the method has not been standardized for determining *T. gondii*. Attempting to use immunomagnetic separation, but indirectly, by means of two monoclonal antibodies (mAbs 3G4, 4B6), seems to have yielded results. The age of the oocysts (1 or 12 months since they are in the environment) does not influence the sensitivity of the results of indirect immunomagnetic separation. This method was also tested by the same author in 2007, confirming the importance of 4B6 monoclonal antibodies in detecting *T. gondii* oocysts in water samples (Dumètre and Dardé 2005, 2007).

Lass et al. (2009) tested 101 soil samples from Poland, by PCR, to determine *T. gondii* oocysts. Soil samples were collected from areas considered cat defecation areas. The oocysts were recovered by flotation, and then subjected to DNA extraction by PCR. *Toxoplasma* DNA could be identified in 18 soil samples, and seven isolates could be genotyped as belonging to genotypes I and II. This study demonstrates, once again, the major role that environmental *T. gondii* oocysts play in the epidemiology of the disease (Lass et al. 2009).

Sotiriadou and Karanis (2008), from the University of Agriculture and Veterinary Medicine of Japan, demonstrated the sensitivity of the loop-mediated isothermal amplification (LAMP) method compared to the PCR or immunofluorescence (IFAT) method to determine *T. gondii* in water. Twenty-six water samples contaminated with a known number of *T. gondii* oocysts were tested. After DNA extraction, the sensitivity of the LAMP method was 100% and that of PCR was 53.8% (Sotiriadou and Karanis 2008).

A total of 52 samples of natural water (from the environment) were tested by LAMP, PCR and IFAT. The LAMP method identified 25 positive samples (48%), the PCR only 7 samples (13.5%) and no positive sample was identified by IFAT. It has been concluded that, by far, LAMP the fastest, most specific and most sensitive method used to determine *T. gondii* from contaminated water (Krasteva et al. 2009).

Winiecka-Krusnell et al. (2009), from the Swedish Institute of Infectious Disease Control, attempted to explain the increased prevalence of *T. gondii* infection in marine mammals. It started from the idea that mollusks have the ability to accumulate oocysts inside them, along with the feeding activity, through filters. It is also known that amoebas are considered potential carriers of pathogens (Winiecka-Krusnell et al. 2009).

Based on these data, the author experienced the ability of amoebas (*Acanthamoeba castellanii*) to accumulate oocysts of *T. gondii*, through their phagocytosis activity. Through this study, it was observed that placed in the same environment, amoebas have the ability to accumulate inside oocysts of *T. gondii*. Overall, 85% of the amoebae contained 1–3 oocysts, and 15% contained more than 3 oocysts. Oral inoculation of these amoebae in mice led to the development of infection, thus demonstrating that amoebas have the ability to transmit *T. gondii* oocysts, preserving their infectivity and pathogenicity. However, the presence of oocysts could not be demonstrated in mollusks, in their natural environment (Winiecka-Krusnell et al. 2009).

It is considered that the possibility of transmitting *T. gondii* to marine mammals through aquatic mollusks exists, but remains, for the time being, at the stage of theoretical hypothesis, and cannot be proven under natural conditions. The role of amoebas in the spread of *T. gondii* oocysts requires further investigation (Murata et al. 2004).

4.3. Routes of Infection for Toxoplasma gondii

The main routes of infection, both in the definitive and in the intermediate hosts, are the oral one with oocysts or through carnivorism (with bradyzoites and tachyzoites) and the congenital one (Mederle et al. 2007). Theoretically, the transmission of toxoplasmosis can also be achieved sexually, through saliva or through the consumption of milk and eggs. The parasitic stages that are most likely involved in these types of transmission would be tachyzoites. They are more sensitive and cannot survive for longer periods outside the host's body. Therefore, the risk of transmission via kissing or venereally is virtually nil. Although *Toxoplasma* have been isolated from ram's sperm, it does not appear that they do not transmit the disease through breeding (Zanetti Lopes et al. 2009). There is a minimal risk of *Toxoplasma* infection following the consumption of cow's milk, especially since it is generally pasteurized, even boiled, but there have been infections due to the consumption of uncooked goat's milk (Sacks et al. 1982; Chiari and Neves 1984; Spišák et al. 2010). Chicken eggs, although important sources for Salmonella infection, are unlikely to transmit *T. gondii* infection.

In recent years, the transmission of toxoplasmosis by organ transplantation or blood transfusion has become important.

Oral transmission of the disease is achieved by consuming feed contaminated with oocysts. In sheep, the parasite can be transmitted through grains that are administered in the last part of pregnancy, but also through the consumption of infected placentas (Ballweber 2006).

In pigs, infection occurs through the consumption of unsterilized kitchen scraps or rodents (Caporali et al. 2005; Dărăbuş et al. 2006).

We consider that, in the transmission of toxoplasmosis, carnivorism has a bigger role than the infection with oocysts from cats. Carnivorism avoids one of the links in the life cycle: the cat. It seems that the most important role in this mode of transmission is played by sheep and pigs. Tissue cysts from meat destined for human consumption are an important source of infection for people. Cysts develop in less than 6–7 days after infection of the intermediate host with oocysts or bradyzoites (Oncel et al. 2005; de Moura et al. 2007; Tekay and Ozbek 2007).

The transplacental route occurs either as a result of a first infection or due to the exacerbation of a chronic form. Tachyzoites cross the placental barrier through tissue necrosis and develop in the pregnant uterus (Anderson and Remington 1974; Canessa et al. 1988; Dunn et al. 1999; Stroia and Ungureanu 2007).

4.4. Favoring Factors in Toxoplasma gondii Infection

Some favoring conditions may increase cats' susceptibility to infection and the deposition of more oocysts. Thus, under experimental conditions, cats infected with *T. gondii* eliminate a greater number of oocysts after ingestion of *Isospora felis*, without causing the disease, suggesting that the infection is limited to the intestine. Under the same conditions, cats chronically infected with *T. gondii*, after administration of corticosteroids, eliminate oocysts, but less than in the case of infection with *I. felis*. However, they clinically manifest the disease. In cats that have eliminated oocysts, intestinal immunity is strong. In these, primary infection with *T. gondii* does not cause immunosuppression (Johnson 1997). Most cats after a primary infection, with elimination of oocysts, and challenge after 6–12 months no longer excrete parasites in the feces. It is understood that infected cats in the first months of life are strongly immunized and after 4–5 months, theoretically, no longer eliminate oocysts, even if, in the case of reinfestations, some phases of *Toxoplasma* development can occur in enterocytes (Durecko et al. 2004; Dărăbuş et al. 2006).

It appears that in about 55% of cats, intestinal immunity persists for up to 6 years. In the past, removal of oocysts after reinfection or re-elimination of oocysts in the absence of reinfection with *T. gondii* was thought to be rare. However, recent studies show that the immunity developed from the primary infection does not persist throughout the cat's life. A new elimination of oocysts can be induced in cats that have been reinfected with *T. gondii* approximately 6 years after primary infection. In addition, in some cases, a re-elimination of oocysts without a cat reinfection was observed for a short time. This is not yet fully understood, but various immunosuppressive causes are assumed, such as over infestations with *Cystoisospora felis* or leukemic virus infections or feline immunodeficiency virus;

factors that induce a re-elimination of oocysts under natural conditions (Airinei et al. 2007; Dubey 2005).

Congenitally infected kittens can also eliminate oocysts. If cats are infected with tachyzoites and/or oocysts, less than 50% of them will eliminate oocysts. Ingestion of bradyzoite cysts ensures 100% elimination of oocysts by infected cats (Dubey et al. 2004c; Miró et al. 2004).

4.5. Methods of Infestation in Toxoplasmosis

During the biological cycle of *T. gondii*, there are three infesting stages: tachyzoites, bradyzoites inside tissue cysts and sporozoites in sporulated oocysts. All three stages are infesting for both hosts, intermediate and final. The ways of infestation are the following (Figure 7):

- Horizontal transmission, by ingesting infesting oocysts from the environment;
- Horizontal transmission, by ingestion of tissue cysts from raw or insufficiently prepared meat or viscera of the intermediate hosts;
- Vertical transmission, by transplacental transmission of tachyzoites.

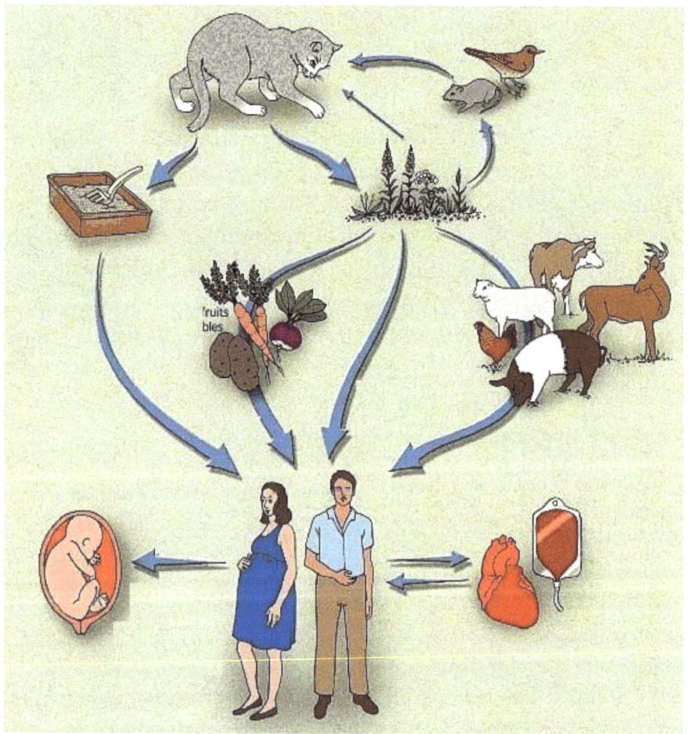

Figure 7. Methods of infestation in toxoplasmosis (Pathways for *Toxoplasma gondii* infection). Source: Authors' compilation based on data from www.aafp.org/afp/2 003/0515/p2131.html (accessed on 28 September 2019).

In addition, in some hosts (goats), tachyzoites can also be transmitted through milk from mother to infant (Dubey 2000; Ishag et al. 2006).

Not all possible routes of infection are of epidemiological importance, and the sources of infection can vary greatly between different ethnic groups and between different geographical areas (Tenter et al. 2000).

In order to prevent horizontal transmission and, in particular, to eliminate the sources of infestation, it is necessary to develop strategies to prevent infection in at-risk groups, in unimmunized pregnant women or in AIDS patients (Cambrea et al. 2007; Colville and Berryhill 2007).

4.5.1. Transmission by Tachyzoites

Tachyzoites play the most important role in vertical transmission. They are very sensitive to environmental factors and are easily destroyed outside the host. Therefore, it is believed that horizontal tachyzoite transmission is not epidemiologically important, although horizontal tachyzoite infections may sporadically occur.

Heart, kidney, liver or bone-marrow transplants can be complicated by *T. gondii* infection, and both tachyzoites and tissue cysts may be involved (Cermáková et al. 2005).

Tachyzoites can also be transmitted through the blood, being found in the contents of white-blood-cell fractions. However, parasitemia only exists for a short time after the first infection. Thus, infection by normal blood transfusions is a minor risk (Béla et al. 2008).

Tachyzoites have also been found in sheep's, goat's and cow's milk, but they cause infections only when unpasteurized goat's milk is consumed (Brandão et al. 2006; Ishag et al. 2006). They are sensitive to proteolytic enzymes and are usually destroyed by gastric digestion. However, Dumètre and Dardé (2003) have shown that tachyzoites can sometimes survive for a short time (>2 h) in pepsin acid solution, and oral administration of high-dose tachyzoites can lead to infections in mice and cats (Dumètre and Dardé 2003).

It has also been suggested that tachyzoites may reach the host by penetrating the mucous membranes and into the blood or lymph circulation before reaching the stomach. This could explain *T. gondii* infection in an infant whose mother acquired a primary infection with this parasite (Robert-Gangneux et al. 1999).

Tachyzoites are sensitive to temperature. This is noteworthy because *T. gondii* infection was observed in two children who frequently drank unpasteurized goat's milk, and their parents, who drank milk only in tea or coffee, remained seronegative. Tachyzoites are destroyed by pasteurization and heating. In Poland, the consumption of unpasteurized milk is considered a risk factor for horizontal transmission (Sacks et al. 1982; Chiari and Neves 1984).

Tachyzoites have been found in other bodily fluids: saliva, sputum, urine, tears and semen, but they are not important in horizontal transmission.

Different studies (Dubey et al. 2005c, 2008a, 2010a) show that tachyzoites can also be found in raw chicken eggs laid by chickens with experimental infection (Dubey et al. 2002b, 2004b, 2004c, 2004d, 2005b, 2006f, 2006b, 2007c). However, marketed eggs should be free of *T. gondii*, and tachyzoites are sensitive to heating and salting, so any method of cooking eggs will destroy these forms of infestation.

The unanimous opinion is that most horizontal transmissions are made by eating meat or organs with cysts and by ingesting oocysts from food or water contaminated with cat feces (Dubey 2004).

4.5.2. Transmission by Tissue Cysts

Meat tissue cysts are an important source of infection for humans. Cysts develop in less than 6–7 days after infection of the intermediate host with oocysts or other cysts.

Among meat animals, cysts are most commonly found in the tissues of infected pigs (Gamble et al. 2005; Schulzig and Fehlhaber 2005; Ştirbu-Teofănescu et al. 2007), sheep and goats (Jittapalapong et al. 2005; Goz et al. 2006; Irabuena et al. 2006; Silva and La Rue 2006), less frequently from infected birds (Dubey et al. 2006e, 2007f), rabbits (Figueroa-Castillo et al. 2006), dogs (Aslantaş et al. 2005; Wanha et al. 2005; Cozma et al. 2007; Silva et al. 2007) or horses (Naves et al. 2005) and very rarely in beef (Yu et al. 2007).

In Europe and the United States, pigs have been considered a major source of human infection (Desmonts and Couvreur 1974). This hypothesis is based on the fact that cysts were found in most pork products (between 1970–1980). Epidemiological studies performed on pig farms with intensive breeding systems in the Netherlands, Austria and Germany showed that the prevalence of *T. gondii* infection in pigs decreased significantly (<1%) (Schulzig and Fehlhaber 2005).

Pigs are often infected by consuming small rodents (dead or sick mice) that carry *Toxoplasma* cysts. In some farms, it has been possible to link the percentage of infected mice to that of pigs positive for *Toxoplasma*. In addition, pigs, under certain circumstances, become cannibals by biting their tails or ears (autophagy) (Oncel and Vural 2006; Cilievici and Creţu 2007).

In several U.S. countries, seroprevalence in older pigs, which are kept on extensive breeding farms and which are more exposed to the environment, is also decreasing (Caporali et al. 2005; Damriyasa and Bauer 2005).

These data demonstrate that it is possible to decrease the risk of *T. gondii* infection in intensive growth systems if the conditions of hygiene, prophylaxis and control are observed. These measures include:

a. Raising farm animals in shelters, throughout their economic life;
b. Shelters should be protected from rodents, birds and insects;
c. The animal feed should be free of microorganisms;
d. Control of access to shelter for food distributors, and for pets not to have access inside the shelters.

Using some of the preventive measures, it is economically possible to raise *T. gondii*-free pigs and poultry, although this has only been achieved in a few countries: the Netherlands, Denmark and Germany (Damriyasa and Bauer 2005).

In contrast, animals kept at pasture will always be associated with *T. gondii* infection due to the fact that cats eliminate oocysts in the environment and the animals have the possibility of permanent contamination (such as sheep and goats with a seroprevalence of 92% and 75%, respectively) (Brandão et al. 2006). This is important because cysts have been found in many sheep and goat tissues, and thus, these animals are sources of infection through both meat and milk.

Seroprevalence is much lower and more varied in horses, rabbits and birds. This could be due to differences in maintenance, hygiene and feed compared to pigs and small ruminants. In dogs, seroprevalence is usually increased because they are always exposed to the natural environment and the cumulative effect over time (Azevedo et al. 2005; Schares et al. 2005).

Tissue cysts of *T. gondii* from deer or other wildlife, including rabbit, wild buffalo, other deer, kangaroos and bears are considered sources of infection for humans (Zarnke et al. 1997; Martinez-Carasco et al. 2004; Dangolla et al. 2006). Hunters and their families can also become infected during the evisceration of animals and the cutting of game.

Due to the frequent consumption of caribou meat in Inuit culture, an association was observed between congenital toxoplasmosis and fur clothes (wolf, fox) in pregnant women (Kelly and Sleeman 2003). These cases show that toxoplasmosis in humans can also occur in the Arctic regions, although infections in these areas are rarer than in regions with hot and humid climates. In addition, there is a cumulative effect over time in wildlife, resulting in an increased prevalence of infection (Omata et al. 2005a, 2005b; Anwar et al. 2006; Ryser-Degiorgis et al. 2006).

In some wild animals, such as marsupials in Australia, *T. gondii* infection did not exist until cats were introduced into their territory (about 100 years ago). As a result, these animals are very susceptible to *T. gondii* infection (Johnson 1996).

Although the seroprevalence of toxoplasmosis is lower in marsupials than in mammals, kangaroo meat has recently been considered a source of infection for humans because it is tender and low in fat and consumed in blood or undercooked (Johnson 1996).

Bradyzoites are more resistant to digestive enzymes (pepsin and trypsin) than tachyzoites. Thus, ingestion of viable tissue cysts by an unimmunized host will cause an infection. Tissue cysts are less resistant to environmental conditions than oocysts, which are relatively resistant to temperature variations, but still cysts remain infectious in carcasses or minced meat at refrigerator temperature (1–4 °C) for over three weeks, probably as long as the meat is good for consumption.

Tissue cysts also survive at −1 to −8 °C for a week but are destroyed at temperatures of −12 °C or lower. However, some can survive even at these temperatures. It has been observed that some lines of *T. gondii* can withstand frost (Meireles et al. 2003).

Instead, meat cysts are destroyed by heating to 67 °C. Survival of cysts at lower temperatures depends on the duration of cooking. For example, under laboratory conditions, the cysts survived at 60 °C for about 4 min and at 50 °C for about 10 min. It should be noted that cooking for a longer time would be necessary to obtain the optimum temperature to kill all cysts in all parts of the meat. Some cysts can remain infectious if the method of cooking the meat is just heating, such as cooking in the microwave.

Some studies (Dubey 2000) have shown that tissue cysts are destroyed by commercial methods of preparing meat, by salting or smoking, at low temperatures.

The survival time of tissue cysts varies depending on the concentration of the saline solution or the freezing temperature. Under laboratory conditions, cysts are destroyed at a concentration of 6% NaCl at all subjected temperatures (4–20 °C) but survive at low concentrations of saline for several weeks. It has also been shown that in home-made pork sausages, the salt applied does not necessarily kill the tissue cysts.

Tissue cysts are destroyed by gamma radiation at a dose of 1.0 KGy, which is supported by US authorities (Dubey 2000). Meat irradiation has been approved in only a few countries and is only possible in developed countries. It is rejected by consumers in many parts of the world.

In 1954, Weinman and Chandler (cited by Dubey 2009a) suggest, for the first time, the possibility of transmitting toxoplasmosis through raw (uncooked) meat (Dubey 2009a). This hypothesis was later tested in a children's sanatorium. They compared the rate of infection in children before and after hospitalization. At first, a rate of 10% was set, then it increased to 50% after the administration of two portions of lightly cooked beef and horse, increasing to 100% after the consumption of undercooked lamb (Dubey 2009a).

Later, the disease was described in a group of medical students who ate a hamburger containing insufficiently cooked meat (Dubey 2009a).

Recent outbreaks of acute toxoplasmosis in humans in different parts of the world show that the sources of infection vary widely among different populations, depending on cultural differences and culinary habits. In Canada, an outbreak of congenital toxoplasmosis in an Inuit colony in northern Quebec has been associated with frequent consumption of caribou meat and fur processing from fur animals, while seropositivity in pregnant women has been associated with the consumption of dried seal meat, seal liver and raw caribou meat (Dubey 2004).

In Australia, the onset of acute and congenital toxoplasmosis was associated with kangaroo and lamb meat that is less cooked during a party in Queensland (Johnson 1996). Meanwhile, at a party in Brazil, the consumption of raw ram meat was incriminated as a source of acute toxoplasmosis in humans (Ueno et al. 2009).

The cysts are located, randomly, in any muscle mass, but especially in the myocardium. In this muscle, in pigs, survival of several years was achieved. It is also important to consider that, in addition to meat, tissue cysts can also form in the visceral organs. Thus, an acute outbreak of toxoplasmosis in humans has been described after the consumption of spleen and raw wild boar liver, and another outbreak after the consumption of raw pork liver in Korea, where the liver is considered a food with special nutritional qualities (Li et al. 2006; Liang et al. 2006). In the second case, both tissue cysts and tachyzoites could be involved.

While these reports show that the risk of *T. gondii* infection through meat or other edible parts of an animal varies with culture and cultural habits in different human populations, data from acute outbreaks of toxoplasmosis are usually related to an occasional source of infection and not a major epidemiological source for the entire population. For example, in Australia, kangaroo meat has only recently been marketed for human consumption (Johnson 1996).

It should be noted that most *T. gondii* infections in immunocompetent people progress asymptomatically, and thus these infections will not be recorded unless systematic programs for recording *T. gondii* infections are applied to these people. Only recently, in a comprehensive control, have studies been carried out specifically to identify the sources of infection in different human populations. In several studies conducted by the European Congenital Toxoplasmosis Research Center, a large number of women have been registered for seroconversion during pregnancy. The European Center has conducted studies (Cook et al. 2000) in various cities in Belgium, Denmark, Italy, Norway, Switzerland and England, identifying the consumption of undercooked lamb, beef or game, soil contact and travel outside Europe and North America, as strong risk factors for *T. gondii* infections, with 30–63% of cases being attributed to the consumption of poorly cooked meat (Cook et al. 2000; Tenter et al. 2000).

Consumption of raw pork and tasting of raw meat during cooking were also considered to be the main risk factors for *T. gondii* infection in the Polish population. Frequent meat consumption or poorly cooked meat consumption have also been associated with seroconversion or seropositivity in *T. gondii* in studies conducted by a routine health check for adults in France, Yugoslavia, and the United States (Cook et al. 2000; Aydenizoz et al. 2005).

While the consumption of raw or undercooked meat has been consistently found to be a risk factor in all these studies, the real importance of this factor, depending on the type of meat, varies in different countries. For example, in France, the consumption of undercooked beef was an important risk factor compared to lamb. In Norway, the consumption of low-cooked lamb was a major factor in infection compared to pork, while in Poland, the consumption of poorly cooked pork was the main risk factor identified during the study. Additionally, in Norway, over 18% of sheep and only 3% of slaughtered pigs had tissue cysts, while in Poland, 36% of slaughtered pigs were infected (Cook et al. 2000; Figliuolo et al. 2004; Negash et al. 2004; Hove et al. 2005b).

These data highlight the differences in the culinary habits of consumers or differences in the prevalence of infection in meat products in the mentioned regions.

To prevent horizontal transmission of *T. gondii* to humans through carrier foods, meat and other edible parts of carcasses should be well cooked (up to 67 °C in the middle of the piece of meat) before being eaten. Although freezing, as the only method of processing, is not a valuable way to destroy all cysts, deep freezing of meat (−12 °C or even lower temperatures) before cooking can reduce the risk of infection. In addition, meat should not be tasted during its preparation or cooking, which is very important for unimmunized pregnant women.

The fact that slaughterhouse species are always at risk of infestation points out that toxoplasmosis in humans has a higher incidence in countries where culinary habits make it possible to eat less cooked meat. Therefore, the incidence of human infection in France is very high, with the French consuming a lot of raw meat. All butchers are possible sources of *Toxoplasma* infection, with pigs having an infection rate of about 70%. In all animals, the density of the infection is not very high: a cyst with bradyzoites per 50–100 g of meat (Garcia et al. 1999; Dumètre and Dardé 2003; Hove et al. 2005b).

In order to reduce the risk of horizontal transmission to humans through tissue cysts, it is essential that preventive measures also include rigorous kitchen hygiene. Thus, in a study conducted in Norway, non-washing of knives each time after cutting raw meat was independently associated with an increased risk of a primary infection during pregnancy (Cook et al. 2000).

Tissue cysts and tachyzoites are destroyed by water and thus, hands and all kitchen utensils used in the preparation of raw meat or other products of animal origin should be thoroughly washed with hot water and detergent.

4.5.3. Transmission by Oocysts

A study in Bombay, India (Dubey 2009a) established the prevalence of *T. gondii* infection in vegetarians as similar to that in the rest of the population.

In 1965, Hutchison (cited by Dubey 2009a) first associated *Toxoplasma* infection with cat feces. Until then, Hutchison had not studied protozoa. He administered tissue cysts with *T. gondii* to cats already infested with *Toxocara cati*. He collected the feces of the cats, left the *Toxocara* eggs in the embryo and then administered them to the mice. Those mice expressed the disease. He repeated the experiment using two cats infested with *T. cati* and two that were free of parasites. *T. gondii* was transmitted only to cats infested with *T. cati*. Thus, Hutchison concluded that *T. gondii* is transmitted through *Toxocara cati* eggs. Hutchison's theory was challenged by Frenkel in 1969 (cited by Dubey 2009a), who demonstrated the presence of *T. gondii* oocysts in the feces of cats that were not infested with nematodes. These data were used to complete the sexual cycle in the cat's small intestine (Dubey 2009a).

After identifying *T. gondii* oocysts, Ben Rachid (1970, quoted by Dubey 2009a) inoculated gundi rodents with oocysts, and the animals died after 6–7 days. Thus, the method of infection of rodent rodents in 1908 is explained. It is assumed that at least one cat existed in Nicolle's research laboratories in Tunisia (Dubey 2009a).

To complete the sexual cycle of the parasite, Dr. J. K. Frenkel tested several species of animals, including wild cats, to eliminate oocysts in captivity. An ocelot, for which he was given meat with tissue cysts, eliminated oocysts of *T. gondii*. Of all the species tested, it was observed that only cats eliminate oocysts. Thus, it was concluded that the sexual phase of the cycle takes place only in the small intestine of cats, resulting in the elimination of oocysts (Dubey 2009a).

T. gondii infections in cats usually develop asymptomatically, and vertical transmission is rare, but latent *T. gondii* infections are common in domestic cats and wild cats worldwide. At least 17 species of wild cats have been found to eliminate oocysts: European and African wild cats, Pallas cat, leopard cat, Amur leopard cat, ocelot, Geoffroy's cat, Pampa's cat, leopard, tiger, cheetah, jaguar, lion, cougar, etc., and in some of them, *T. gondii* infections have been shown serologically (Dubey 2005; Mucker et al. 2006).

The degree of seropositivity (by ELISA) increases with age: 22% in cats under one year and 80% in those over 10 years. Between 9 and 46% of pet cats in Europe and the USA show past exposure to the parasite in a serological test, while the seroprevalence of toxoplasmosis in Asia has been estimated at between 6 and 9% (Dubey 2005).

The cat becomes an oocyst eliminator only after it has begun to receive raw meat or become a predator. This means that he must be over one-and-a-half-months old. In addition, up to this age, kittens benefit from passive maternal immunity through Ig G antibodies that have been transmitted to them through the placenta and colostrum. In domestic cats, anti-*Toxoplasma* antibodies can be detected in over 74% of the adult cat population, depending on the type of food and the environment in which the cats are kept: indoors or outdoors (in the yard). Seroprevalence is usually higher in stray and wild cats than in urban and suburban settlements. After primary infection with *T. gondii*, cats that are kept indoors can eliminate a large number of oocysts inside the home, thus exposing their owners to the risk of infection. Stray cats and cats that are living near the farms or in farms can contaminate the environment with oocysts within a radius of about 1.5–3.5 km and thus to infect the animals that will then be slaughtered for human consumption (Asthana et al. 2006; Lazăr and Barbu 2007).

Domestic cats and other feline species can be infected with *T. gondii* by ingesting infectious oocysts from the environment or by ingesting tissue cysts from intermediate hosts (Dumètre and Dardé 2003). For example, cats can ingest tissue cysts by eating food scraps containing meat or beef or game organs. Cats that hunt can become infected through the consumption of small mammals or birds infected with *T. gondii*. Depending on the host species, the geographical territory and the season, over 73% of small rodents and over 71% of wild birds can be infected (Dubey 2002; Djurković-Djaković et al. 2005).

After the primary infection of cats with tissue cysts, in the small intestine, bradyzoites immediately begin a phase of asexual multiplication which consists of numerous cycles of endopolygeny. The terminal stages of this multiplication begin the sexual phase of multiplication, which ends with the formation of oocysts. Almost all cats that were primarily infected with tissue cysts eliminated oocysts after a pre-existing period of 3–10 days, which ended after 20 days.

In contrast, after infection with *T. gondii* oocysts, sporozoites first begin an asexual multiplication phase that is similar to the development of the intermediate host in the extraintestinal tissues. During this development, some parasites migrate to the intestinal tissues and begin a sexual phase of multiplication. One-third of cats that have been primarily infected with oocysts eliminate other oocysts for about 10 days after a prepatent period longer than 18–49 days.

Cats can also become infected by ingesting a large number ($\geq$1000) of tachyzoites, which will lead to the elimination of oocysts for about 7 days, after 15–19 days after ingestion of tachyzoites (Dubey 2005).

Cats eliminate oocysts in large numbers only after a primary *T. gondii* infection. In the past, the removal of oocysts after a reinfection or a new elimination of oocysts, in the absence of a reinfection with *T. gondii*, was thought to be rare. However, some studies (Dubey et al. 2004c) show that immunity developed from primary infection does not persist throughout the cat's life (Miró et al. 2004; Michalski and Platt-Samoraj 2004; Dubey et al. 2007g). A new elimination of oocysts may be induced in cats that have been reinfected with *T. gondii* approximately 6 years after primary infection. Experimentally, the elimination of oocysts can be induced by superinfection with other coccidia, for example with species of the genus *Isospora*, as well as by immunosuppression, for example following an increased dose of corticosteroids.

The domestic cat is the only domestic animal that is a definitive host for *T. gondii* and appears to play a key role in the epidemiology of infection. At the time of elimination by the cat, the oocysts are sporulated and are not immediately infectious. Thus, direct contact with cats does not result in a *T. gondii* infection. Keeping cats and keeping them inside houses or apartments does not necessarily pose a risk of infection if preventive measures are strictly followed and cat feces are removed from the house daily.

Sporulated oocysts in the environment are a potential source of infection for humans and other intermediate hosts. The epidemiological importance of oocysts is demonstrated by the fact that, despite the worldwide spread of *T. gondii*, infections with this parasite are completely absent on small islands and atolls that have never been populated by cats. Contamination of the environment with *T. gondii* oocysts is to infect other domestic cats or wild cats (Wallace 1969; Mucker et al. 2006).

After primary infection with tissue cysts or *T. gondii* oocysts, a single cat can eliminate more than 100 million oocysts in the environment. In favorable environmental conditions, with oxygen, humidity and high temperatures, the oocysts sporulate and become infectious in 1–5 days. The sporulation period can be delayed by unfavorable environmental conditions.

Depending on the pathogenicity of *T. gondii*, ingestion of 10 sporulated oocysts can cause infection in intermediate hosts, and ingestion of over 100 sporulated oocysts causes feline infection, thus resuming soil contamination with oocysts (Miró et al. 2004; Hove and Mukaratirwa 2005; Venturini et al. 2004; Xie et al. 2005).

Sporulated oocysts are very resistant in the environment. They survive poorly in low temperatures and dehydration but remain infectious in moist soil and sand for over 18 months. Under laboratory conditions, they survive at 4 °C for over 54 months, and frozen at −10 °C, for 106 days, but they are killed in 1–2 min at 55–60 °C. They are waterproof and therefore particularly resistant to disinfectants (Dumètre and Dardé 2003). Ordinary disinfectants have no effect on sporulated oocysts, at least when used in usual concentrations, except for 10% ammonia. Fermentation and putrefaction, which occur at high temperatures and create anaerobiosis, destroy oocysts. By destroying the germs of putrefaction and fermentation, common disinfectants prolong the survival of oocysts.

Cat—The "Key" Element in the Transmission of *T. gondii*

There are several preventative measures that could reduce the risk of horizontal transmission of *T. gondii* to humans through oocysts.

Regarding the spread of oocysts by pet cats, the owners of cats belonging to risk groups including non-immune pregnant women or immunocompromised patients should be advised and warned.

Fecal examination is not a sufficient measure, as most cats eliminate oocysts in just 1–2 days, while the entire prepatent period can take more than 20 days (Airinei et al. 2007; Afonso et al. 2008). In those days when only a few oocysts are removed, and in case of re-eliminations, the coproscopic methods can give negative results, although the feces are infesting for the intermediate hosts, which is confirmed by mouse inoculation. Thus, fecal examination is relevant for cat owners only if the result is positive, if *T. gondii* oocysts have been identified in cat feces (Salant and Spira 2004).

For this reason, it is more appropriate to perform a serological examination in cats to identify anti-*Toxoplasma* antibodies (Ig G) (Dubey et al. 2004c). A negative serological result shows that the cat has not yet been exposed to *T. gondii* and is susceptible to future infections. In such cases, cats should be fed commercial (canned or dried) or cooked food (cooked meat) and should not be allowed to hunt to prevent a primary *T. gondii* infection. In addition, the environment in which the cat lives must be controlled to exclude intermediate hosts (mice and rats), and also vectors (kitchen beetles or other invertebrates) (Chinchilla et al. 1994).

A positive serological result shows that the cat has already been infected with *T. gondii* in the past. Most cats with a detectable level of Ig G are probably immune and will not eliminate oocysts in the near future. Cats that have been infested with tissue cysts usually become positive between the second and fifth week post infestation. In some cats, which have been infested by ingestion of oocysts, Ig G antibodies are already detectable during the prolongation of the prepatent period (Dubey 2005).

Thus, although the identification of Ig G antibodies in cats' serum is, in most cases, an indicator of immunity, this does not rule out the possibility that, in some rare cases, cats may be able to eliminate oocysts. In addition, some cats that have been infected with *T. gondii* in the past may re-eliminate oocysts for a short period of time, so immunosuppressive treatments of cat owners belonging to risk groups should be avoided. It would also be advisable for immunocompromised individuals not to keep cats in their homes.

However, the feces of pet cats should be removed from the house every day. Cat litter boxes and all objects that may come in contact with feces should be washed thoroughly with hot water (above 70 °C) and detergent by someone wearing gloves, but preferably not by immunocompromised individuals or by pregnant women (Dumètre and Dardé 2003).

On the other hand, although the consumption of poorly cooked meat has been found to be the main risk factor in several recent studies of *T. gondii* infections in humans, these findings do not explain the high rate of seropositivity (24–47%) in some vegetarian populations (Roghmann et al. 1999).

Some of the risk factors identified in these studies point to the importance of oocysts in the transmission of the disease to humans. For example, soil contact was found to be a strong risk factor in the European Toxoplasmosis Control Center study, and 6–7% of primary human infections were attributed to this factor, while in a study in Norway, the consumption of unwashed raw fruits and vegetables was considered an important factor of primary infections during pregnancy (Cook et al. 2000).

It is recommended that pregnant women and immunocompromised individuals wash or heat-cook vegetables and fruits that may be contaminated with cat feces. These individuals should also wear gloves when working in gardens.

In addition, *T. gondii* oocysts in children's playgrounds can be a source of infection. Geophagy was strongly associated with an outbreak of acute toxoplasmosis in 6 of the 11 preschool children in a large family who played in the same playground in their grandparents' yard—a pit to which cats also had access. Preventive measures could be taken to prevent cats from defecating in children's playgrounds by covering sand pits when not in use (Dubey 2000; Dumètre and Dardé 2003; Chen et al. 2005).

Three other outbreaks of acute toxoplasmosis in humans have been associated with soil contamination with oocysts. In 1977, an outbreak of the disease in 37 of 86 customers of a riding farm in Atlanta, USA, was associated with the inhalation of oocysts from the air—oocysts that were eliminated by farm cats (Naves et al. 2005).

Another occurrence of toxoplasmosis in 35 of 98 military instructors was associated with ingestion of water contaminated with oocysts during training on swampy terrain in Panama.

The largest and most studied acute toxoplasmosis outbreak in humans occurred in 110 individuals in Vancouver, Canada, in 1995 (Cook et al. 2000). Retrospective epidemiological studies provide important evidence that this outbreak was caused by contamination of drinking water with oocysts. Both domestic cats and wild cats (cougars) have been shown to be the source of infection through oocyst contamination of the reservoir, which was the main source of drinking water into which unfiltered surface water is discharged (Dubey 2004; Mucker et al. 2006).

4.6. The Prevalence of Toxoplasma gondii Infection in Animals

4.6.1. *T. gondii* Infection in Livestock

Sheep and goats

Toxoplasma gondii is one of the common causes of abortions in sheep and goats, and thus a major problem in the production of lambs and goats. If several abortions occur on a farm, it means that the animals have access to an important source of oocysts (Figure 8). Sheep and goats develop very strong immunity after an abortion. Vaccination against *Toxoplasma*-induced abortion is practiced in some countries. The diagnosis of abortion can be made on the basis of antibodies from fetal fluids, by the modified agglutination test. Insufficiently cooked lamb, sheep or goat meat is an important source of infection in humans (Weiss and Kim 2004; Dubey et al. 2014a, 2008d; Dubey 2009b).

Figure 8. Free access of cats in sheep farms. Source: Photos by authors.

Mainar-Jaime and Barberán (2007) tested the efficacy of two methods of serological diagnosis to determine the prevalence of *T. gondii* infection in Spanish sheep. Two-hundred and three sheep were tested and a prevalence of 40.4% by MAT and

36.9% by ELISA was obtained (Mainar-Jaime and Barberán 2007). Through comparison with other studies (Garcia et al. 2008) that tested the same methods, but on different species, it can be pointed out that the accuracy of the tests differs depending on the species being tested.

Shaapan et al. (2008) tested the efficacy of four serological diagnostic methods in *T. gondii* infection in sheep. Three hundred samples were collected from sheep in Cairo, Egypt, obtaining different prevalence depending on the method applied: 43.7% by MAT, 41.7% by ELISA, 37% by IFAT and 34% by SFDT. Through comparison with SFDT, which is considered the reference method in toxoplasmosis, different sensitivities and specificities were obtained, as follows: the most sensitive method being MAT, followed by ELISA and then IFAT, and the most specific method being IFAT, followed by MAT and then ELISA. It was concluded that MAT is the most sensitive method of diagnosis in toxoplasmosis, followed by ELISA—with a high sensitivity, which differs depending on the antigen with which the plaque is lined. ELISA also has the advantage of being a quantitative method. IFAT has high specificity but has lower sensitivity (Shaapan et al. 2008).

Silva et al. (2009) revealed infections with *Hammondia heydorni*, *Neospora caninum* and *Toxoplasma gondii* in sheep in Bahia, Brazil. Samples from 102 sheep were examined by PCR, identifying a prevalence of 3.92% for *H. heydorni*, 1.96% for *N. caninum* and 7.84% for *T. gondii*. These results are low, probably due to the fact that Bahia is a dry area of Brazil, an area unfavorable to the sporulation and survival of oocysts in the soil (Silva et al. 2009).

Lopes et al. (2009) tested the effects of *T. gondii* infection on the quality of semen in rams in Sao Paulo, Brazil. He divided eight rams into three groups. Group I received 2×10^5 oocysts, group II—1×10^6 tachyzoites, and group III represented the control group. After inoculation, some clinical signs began to appear, such as hypothermia and anorexia. All rams had anti-*Toxoplasma* antibodies. The semen was followed by volume, type of movements, speed of movement, concentration and morphology of sperm.

Slight changes in sperm quality were observed, but they could not be directly attributed to *Toxoplasma* infection because some changes were observed, including in the control group. Although the experimental infection caused clinical signs, it did not contribute to the alteration of semen in the rams studied (Lopes et al. 2009).

In Brazil, Faria et al. (2007) tested the prevalence of *T. gondii* infection in goats. Three-hundred and six serum samples were tested using an IFAT technique, and a prevalence of 24.5% for *T. gondii* and 3.3% for *Neospora caninum* was identified. The seroprevalence of toxoplasmosis by sex was 30.8% in males and 20.4% in females. This study confirms the increased prevalence of *T. gondii* infection in goats, described by other authors (Faria et al. 2007).

In addition to the data described above, the prevalence of *T. gondii* infection in small ruminants has also been studied by well-known authors in the field in other countries (Table 5).

Table 5. Seroprevalence of *T. gondii* infection in sheep and goats.

(Author Name/Year)	Location	Species	Examined Samples (No.)	Positive Samples (%)	Methods
(Almeida et al. 2021)	Portugal/ Serra da Estrela region	Sheep (farms)	168	69.1%	ELISA
(Gazzonis et al. 2020)	Italy/ Lombardy	Sheep (abattoirs)	227	28.6%	ELISA
		Goats (abattoirs)	51	27.5%	
(Jiménez-Martín et al. 2020)	Spain/ Southern	Sheep (farms)	998	46.5%	MAT
(Pagmadulam et al. 2020)	Mongolia	Sheep (farms)	882	34.8%	ELISA
		Goats (farms)	1078	32%	
(Esubalew et al. 2020)	Ethiopia	Sheep	295	76%	LAT
		Goats	281	65%	
(Tagel et al. 2019)	Estonia	Sheep	1599	41.7%	DAT
(Rouatbi et al. 2019)	Tunisia	Rams (farms)	92	39.1%	ELISA
(Dahourou et al. 2019)	Senegal/ Dakar	Sheep (farms)	278	60.1%	MAT
(Al Hamada et al. 2019)	Iraq/Dohuk	Sheep (farms)	335	42.1%	LAT
		Goats (farms)	97	36.1%	
(Sah et al. 2018)	Bangladesh	Sheep	552	11.96%	ELISA
		Goats	300	16%	
(Dahmani et al. 2018)	Algeria/El Harrach	Sheep (abattoirs)	580	8.3%	ELISA
(Tilahun et al. 2018)	Ethiopia/ Oromia	Sheep (farms)	332	33.7%	ELISA
		Goats (farms)	410	27.5%	
(Moskwa et al. 2018)	Poland/ Kosewo Górne	Sheep (farms)	64	47%	ELISA
		Goats (farms)	39	21%	

(Author Name/Year)	Location	Species	Examined Samples (No.)	Positive Samples (%)	Methods
(Deksne et al. 2017)	Latvia	Sheep	1039	17.2%	ELISA
(Bártová et al. 2017a)	Czech Republic/ Lány	Sheep	24	25%	ELISA
(Elfadaly et al. 2017)	Egypt	Sheep (abattoirs)	254	64.2%	ELISA
		Goats (abattoirs)	293	43.3%	
(Liassides et al. 2016)	Cyprus	Sheep (farms)	515	44.9%	ELISA
		Goats (farms)	581	35.8%	
(Zhou et al. 2017)	Turkey	Sheep	610	20%	ELISA
		Goats	249	12.9%	
(Satbige et al. 2016)	India/ Chennai	Sheep (abattoirs)	136	30.1%	MAT
		Goats (abattoirs)	57	28%	
(Sadek et al. 2015)	Egypt/Assiut	Sheep (farms)	58	39.6%	LAT
		Goats (farms)	47	38.3%	
(Gazzonis et al. 2015)	Italy/ Lombardy	Sheep (farms)	502	59.3%	IFA
		Goats (farms)	474	41.7%	
(Holec-Gąsior et al. 2015)	Poland/ Pomerania	Sheep (farms)	1646	55.9%	ELISA
(Paştiu et al. 2015a, 2015b)	Romania/ Central	Goats (abattoirs)	181	33.1%	ELISA
(Nematollahi et al. 2014)	Iran/ Azarbaijan	Sheep	186	6.9%	ELISA
(Gebremedhin and Gizaw 2014)	Ethiopia	Sheep (farms)	124	31.4%	ELISA
		Goats (farms)	60	15%	

Table 5. Cont.

(Author Name/Year)	Location	Species	Examinated Samples (No.)	Positive Samples (%)	Methods
(Gebremedhin et al. 2014)	Ethiopia/ Shewa	Sheep (abattoirs)	305	20%	MAT
(Leblebicier and Yıldız 2014)	Turkey/ Silopi	Sheep (farms)	100	97%	IFA
(Djokic et al. 2014)	Serbia/ Nationwide	Goats (farms)	431	73.3%	MAT
(García-Bocanegra et al. 2013)	Spain/ Seville	Sheep (farms)	503	49.3	ELISA
(Sechi et al. 2013)	Italy/Tuscany	Sheep (farms)	630	33.9%	IFA
(Glor et al. 2013)	Switzerland/ Zurich	Sheep (abattoirs)	96	18.7%	ELISA
(Diakoua et al. 2013)	Greece	Sheep (farms)	458	53.7%	ELISA
		Goats (farms)	375	61.3%	
(Alvarado-Esquivel et al. 2012a)	Mexico/ Durango	Sheep (farms)	511	15.1	MAT
		Goats (farms)	492	32.7%	
(Balea et al. 2013)	Romania/Cluj County	Sheep (farms)	239	57.7%	ELISA
(Tzanidakis et al. 2012)	Greece/ Northern	Sheep (farms)	1501	48.6%	ELISA
		Goats (farms)	541	30.7%	
(Hotea et al. 2012)	Romania/ Caras-Severin County	Sheep	450	61.33%	ELISA
(Iovu et al. 2012)	Romania/ Transilvania	Goats (dairy farms)	735	52.8%	ELISA
(Bartova and Sedlak 2012)	Czech Republic	Goats (farms)	251	66.6	ELISA
(Hotea et al. 2011a)	Romania/ Timis County	Goats (female)	250	71.6%	ELISA
		Goats (male)	250	78.4%	

Table 5. *Cont.*

(Author Name/Year)	Location	Species	Examined Samples (No.)	Positive Samples (%)	Methods
(Hotea et al. 2011a, 2011b)	Romania/ Timis County	Lambs	200	6.5%	ELISA
		Rams	750	65.73%	
		Sheep	900	36.33%	
(Berger-Schoch et al. 2011)	Switzerland	Sheep	250	2%	PCR
(Czopowicz et al. 2011)	Poland	Goats	1060	30.2%	ELISA
(Camossi et al. 2011)	Brazil/Sao Paulo	Sheep (dairy)	139	5.03%	PCR
(Langoni et al. 2011)	Brazil/Sao Paulo	Sheep	382	18.6%	MAT
(Opsteegh et al. 2010)	Netherlands/ Bilthoven	Sheep	1179	27.8%	ELISA
(Jokelainen et al. 2010)	Finland/ Helsinki	Sheep	1940	24.6%	DAT
(Lopes et al. 2010)	Brazil/Sao Paulo	Sheep	488	52.0%	IFAT
(Mason et al. 2010)	UK/Leeds	Sheep (newborn lambs)	507	9.0%	PCR
		Sheep (lambs at 4 months)	746	7.6%	
(Panadero et al. 2010)	Spain/Galicia	Sheep	177	57%	DAT
(Hotea et al. 2009)	Romania/Arad County	Sheep	400	42.5%	ELISA
(Bártová et al. 2009a)	Czech Republic	Sheep	547	59%	ELISA
(Silva et al. 2009)	Brazil/Bahia	Goats	102	7.84%	PCR
(Carneiro et al. 2009a)	Brazil/Minas Gerais	Sheep	711	31.1%	ELISA
				43.2%	IFAT
(Spilovská et al. 2009)	Slovak Republic/Kosice	Sheep	382	24.3%	ELISA
(Carneiro et al. 2009b)	Brazil/Belo Horizonte	Goats	767	43.0%	ELISA
				46.0%	IFAT

Table 5. Cont.

(Author Name/Year)	Location	Species	Examined Samples (No.)	Positive Samples (%)	Methods
(Ramzan et al. 2009)	Pakistan/ Rahim Yar Kham	Sheep	100	11.2%	LAT
		Goats	100	25.4%	
(Dubey et al. 2008a)	USA	Sheep (lambs)	383	27.1%	MAT
(Górecki et al. 2008)	Poland	Sheep	41	53.65%	IFAT
(Ragozo et al. 2008)	Bulgaria	Sheep	495	24.2%	MAT
(Ueno et al. 2009)	Brazil/Central region	Sheep	1028	38.22%	IFAT
(Soares et al. 2009)	Brazil/Rio Grande do Norte	Sheep	409	20.7%	IFAT
(Neto et al. 2008)	Brazil/Rio Grande do Norte	Goats	306	30.6%	IFAT
(Fusco et al. 2007)	Italy/ Campania	Sheep	1170	28.5%	IFAT
		Sheep (dairy)	1170	0.34%	PCR
(Clementino et al. 2007)	Brazil/Lajes	Sheep	102	29.41%	ELISA
(Vesco et al. 2007)	Italy/Sicily	Sheep	1876	49.9%	ELISA
(de Moura et al. 2007)	Brazil/Parana	Sheep	157	7%	IFAT
(Stimbirys et al. 2007)	Lithuania	Sheep	354	42.1%	ELISA
(Teshale et al. 2007)	Ethiopia	Goats	641	74.8%	MAT
(Hove et al. 2005a)	Zimbabwe	Sheep and Goats	335	67.9%	IFAT
(Klun et al. 2007)	Serbia	Sheep	511	84.5%	MAT
(Goz et al. 2006)	Turkey/Hakari	Goats	92	7.6%	IHA
(Irabuena et al. 2006)	Uruguay	Sheep	239	18%	LAT and ELISA
(Silva and La Rue 2006)	Brazil	Sheep (lambs ≤ 3 months)	120	18.3%	Serology

Table 5. Cont.

(Author Name/Year)	Location	Species	Examinated Samples (No.)	Positive Samples (%)	Methods
(Morshedi et al. 2006)	Iran/Urmia	Sheep	110	21.8%	ELISA and IFAT
(Dumètre et al. 2006)	France/ Haute-Vienne	Sheep (lambs)	164	22%	DAT
		Sheep	93	65.6%	
(Oncel and Vural 2006)	Turkey /Istanbul	Sheep	181	31%	ELISA
(Borde et al. 2006)	Tobago	Goats	161	44%	LAT
(Figliuolo et al. 2004)	Brazil/Sao Paulo	Sheep	597	34.7%	IFAT
		Goats	394	28.7%	
(Ogawa et al. 2005)	Brazil/Parana	Sheep	339	54.6%	IFAT
(Negash et al. 2004)	Nazareth/ Ethiopia	Sheep	116	52.6%;	MDAT
				56%	ELISA
		Goats	58	24%;	MDAT
				25.9%	ELISA
(Belbacha et al. 2004)	Morocco/ Marrakech	Sheep (brain)	50	30%	Histological
(Jittapalapong et al. 2005)	Thailand/Satun	Goats	631	27.9%	LAT
(Meireles et al. 2003)	Brazil/Sao Paulo	Sheep	200	31%	ELISA
		Goats	200	17%	
(Sevgili et al. 2005)	Turkey/ Sanhurfa	Sheep	300	55.66%	SFDT
(Aydenizoz et al. 2005)	Turkey/ Kirikkale	Goats	137	48.9%	IFAT
(Oncel et al. 2005)	Turkey/Yalova	Sheep	63	66.6%	SFDT
				65.08%	LAT
(Sawadogo et al. 2005)	Morocco/ Marrakech	Sheep	261	27.6%	ELISA
(Pereira-Bueno et al. 2004)	Spain	Sheep (who aborted)	173	23.1%	ELISA, IFAT, histopato-logical and PCR
		Sheep (abortions)	106	34.9%	

Table 5. *Cont.*

(Author Name/Year)	Location	Species	Examined Samples (No.)	Positive Samples (%)	Methods
(Demissie and Tilahun 2004)	Ethiopia	Sheep	375	34%	MDAT
		Goats	93	35%	
(Michalski and Samoraj 2004)	Poland	Sheep	20	55%	ELISA
		Goats	142	63.6%	
(Masala et al. 2003)	Italy/Sardinia	Sheep	5787	28.4%	IFAT
		Goats	3852	12.3%	
(Mainardi et al. 2003)	Brazil/Sao Paulo	Goats	442	14.5%	IFAT
(El-Moukdad 2002)	Syria	Sheep	810	44.56%	Serology
(Figueiredo et al. 2001)	Brazil/ Uberlandia	Goats	174	19%	IHA
				19.5%	IFAT
				19.5%	ELISA
(Bisson et al. 2000)	Africa/Uganda	Goats	784	30.61%	IFAT
(Gorman et al. 1999)	Chile	Sheep	408	12%	IAHT
				28%	IFAT
(Hejlicek and Literak 1994a)	Czech Republic	Sheep	886	55%	DT
(Kovacova 1993)	Slovakia	Sheep	1939	10%	DT

Source: Table by authors.

Pigs

Toxoplasma gondii can cause neonatal death in pigs, but *Toxoplasma* abortion is rare in this species. The diagnosis of abortion can be made in sows based on antibodies from fetal fluids via the modified agglutination test. Pigs kept in poor conditions are much more exposed to parasitism (Figure 9) (Table 6). Insufficiently cooked pork is an important source of infection in humans, as tissue cysts can remain viable in animals for more than 865 days (Weiss and Kim 2004).

Figure 9. Pigs breeding in households. Source: Photos by authors.

Table 6. Seroprevalence of *T. gondii* infection in pigs.

(Author Name/Year)	Location	Species	Examinated Samples (No.)	Positive Samples (%)	Methods
(Thaw et al. 2021)	Myanmar	Pigs (Backyard)	256	18.4%	LAT
(Sroka et al. 2020)	Poland/ Nationwide	Pigs	3111	11.9%	MAT
(Zhang et al. 2020)	China/ Shanghai	Pigs (Abattoir)	1158	13.8%	ELISA
(Gisbert Algaba et al. 2020)	Belgium/ Wallonia	Pigs (Abattoir)	92	16.3%	MAT
(Paştiu et al. 2019)	Romania/ Transylvania	Pigs (Backyard)	94	46.8%	IFAT
(Thakur et al. 2019)	India/ Chandigarh	Pigs (Abattoir)	151	48.3%	ELISA
(Felin et al. 2019)	Finland	Pigs	1116	1%	ELISA
(Pablos-Tanarro et al. 2018)	Spain/ Southwestern	Pigs (farms)	2492	8.9%	MAT
(Pipia et al. 2018)	Italy/Sardinia	Pigs (farms)	414	51.7%	ELISA
(Hou et al. 2018)	China/Jiangsu Province	Pigs	141	46.81%	PCR
(Kofoed et al. 2017)	Denmark	Pigs	254	15%	ELISA
(Feitosa et al. 2017)	Brazilia	Pigs	120	12.5%	IFAT
(Tuda et al. 2017)	Indonesia/ North Sulawesi	Pigs (farms)	310	2.3%	LAT
(Kuruca et al. 2017)	Serbia/ Northern	Pigs (Abattoir)	182	17%	MAT

Table 6. Cont.

(Author Name/Year)	Location	Species	Examinated Samples (No.)	Positive Samples (%)	Methods
(Wallander et al. 2016)	Sweden	Pigs	975	5.7%	ELISA
(Papatsiros et al. 2016)	Greece/ Mainland	Pigs (farms)	609	4.3%	ELISA
(Herrero et al. 2016)	Spain/Aragón	Pigs (fattening)	1200	24.5%	IFAT
(Powell et al. 2016)	United Kingdom	Pigs (Abattoir)	620	7.4%	Dye Test
(Felin et al. 2015)	Finland/ Nationwide	Pigs (Abattoir)	1353	3.2%	ELISA
(Alvarado-Esquivel et al. 2015)	Mexico/Baja California Sur	Pigs (Abattoir)	308	13%	MAT
(Onyiche and Ademola 2015)	Nigeria/ Ibadan	Pigs (Abattoir)	302	29.1%	ELISA
(Matsuo et al. 2014)	Japan/Gifu	Pigs (Abattoir)	155	5.2%	LAT
(Esteves et al. 2014)	Portugal/ Lisbon	Pigs (Abattoir)	381	7.1%	MAT
(Hernández et al. 2014)	Spain/ Andalusia	Pigs (free range)	709	27.1%	ELISA
(Devleesschauwer et al. 2013)	Nepal/ Kathmandu	Pigs (Abattoir)	742	11.7%	ELISA
(Paştiu et al. 2013)	Romania/ Center, West, Northwest	Sows	371	12.4%	IFAT
(Turčeková et al. 2013)	Slovak Republic/6 regions	Pigs (Abattoir)	923	4.2%	ELISA
(Basso et al. 2013)	Argentina	Outdoor pigs	149	80.5%	IFAT
				83%	Western blot
				83.2%	ELISA
		Indoor pigs	148	8.1%	IFAT
				23%	Western blot
				8.8%	ELISA

(Author Name/Year)	Location	Species	Examined Samples (No.)	Positive Samples (%)	Methods
(Balea et al. 2013)	Romania/Cluj	Pigs (Backyard)	187	43.1%	ELISA
(Muñoz-Zanzi et al. 2012)	Chile/ Southern	Pigs (Abattoir)	340	8.8%	ELISA
(Alvarado-Esquivel et al. 2012b)	Mexico/ Oaxaca	Pigs (Backyards)	337	17.2%	MAT
(Hotea et al. 2011a)	Romania/ Caras-Severin County	Pigs (extensive system)	950	23.79%	ELISA
(Hotea et al. 2011a)	Romania/Arad County	Pigs (extensive system)	700	16.28%	ELISA
(Hotea et al. 2011a)	Romania/ Timis County	Pigs (extensive system)	1600	21.93%	ELISA
		Pigs (semi-intensive system)	400	1.5%	
		Pigs (intensive system)	1700	0.88%	
(Frazão-Teixeira et al. 2011)	Brazil/ Rio de Janeiro	Pigs (heart)	16	12.5%	PCR
		Pigs (brain)	19	15.78%	
(Berger-Schoch et al. 2011)	Switzerland	Pigs	270	2.2%	PCR
(García-Bocanegra et al. 2010a)	Spain	Sows	400	24.2%	MAT
		Pigs (fattening)	1570	9.7%	
(García-Bocanegra et al. 2010b)	Spain/ Catalonia	Pigs	1202	19%	MAT
(García-Bocanegra et al. 2010c)	Spain	Piglets (1–25 weeks)	73	35.6%	MAT
(Huang et al. 2010)	China/Fujian	Sows	605	14.38%	IHAT
(Zou et al. 2009)	China/ Yunnan	Pigs	831	16.97%	IHAT

(Author Name/Year)	Location	Species	Examined Samples (No.)	Positive Samples (%)	Methods
(Villari et al. 2009)	Italy/Sicily	Pigs (5–7 months)	877	1.1%	ELISA
		Pigs ($\geq$11 months)	125	8.9%	
(Correa et al. 2008)	Spain/ Panama	Pigs	290	32.1%	IFAT
(Sroka et al. 2008)	Poland	Pigs	106	26.4%	MAT
(Srikitjakarn et al. 2008)	Thailand	Pigs (farms)	322	7%	LAT
(de Moura et al. 2007)	Brazil/ Parana	Pigs	117	8.54%	IFAT
(Van der Giessen et al. 2007)	Netherlands	Pigs (intensive system)	265	0.38%	ELISA
		Pigs (extensive system)	178	5.62%	
		Pigs (from households)	402	4–33%	
(Belfort-Neto et al. 2007)	Brazil/ Erechim	Pigs (diaphragm)	50	34%	PCR
		Pigs (tongue)	50	66%	
(Huong and Dubey 2007)	Vietnam	Pigs	587	27.2%	MAT
(Caporali et al. 2005)	Brazil/ Sao Paulo/ Pernambuco	Pigs	759	66.67%	IFAT and MAT
(Cavalcante et al. 2006)	Brazil/ Rondonia	Pigs	80	37.5%	MAT
				43.7%	IFAT
(Hill et al. 2006)	USA	Pigs	69	56.52%	ELISA
(de Sousa et al. 2006)	Portugal	Pigs	333	15.6%	MAT
(Thiptara et al. 2006)	Thailand	Piglets (17 days old)	14	71.43%	LAT
(Klun et al. 2006)	Serbia	Pigs	605	28.9%	MAT

Table 6. Cont.

(Author Name/Year)	Location	Species	Examined Samples (No.)	Positive Samples (%)	Methods
(Schulzig and Fehlhaber 2005)	Germany	Pigs	100	5.6%	Serology
(Hove et al. 2005b)	Zimbabwe	Pigs	474	35.71%	IFAT
(Damriyasa and Bauer 2005)	Germany/Munsterland	Sows (farms)	1500	9.3%	ELISA
(Gamble et al. 2005)	USA	Pigs	70	85.7%	MAT
				88.6%	ELISA
(de A Dos Santos et al. 2005)	Brazil/Sao Paulo	Pigs	286	17%	MAT
(Venturini et al. 2004)	Argentine	Sows	230	37.8%	MAT
(Tsutsui et al. 2004)	Brazil/Parana	Pigs	521	15.35%	IFAT
(Fialho and de Araujo 2002)	Brazil/Porto Alegre	Pigs	240	20%	IFAT
				33.75%	IHA
(Saavedra and Ortega 2004)	Peru/Lima	Pigs	137	27.7%	Western bloot
	USA/Georgia		152	16.4%	
(Damriyasa et al. 2004)	Germany	Sows	2041	19%	ELISA
(Fan et al. 2004)	Taiwan	Pigs	111	28.8%	LAT
(Silva et al. 2003)	Brazil	Pigs	115	86.08%	MAT
(Fehlhaber et al. 2003)	Germany	Pigs	1005	20.5%	ELISA
(Ortega et al. 2003)	USA/Peru	Pigs	137	28%	IB
(Dubey et al. 2002c)	Massachusetts	Pigs	55	92.72%	Inoculation on cats
(Inoue et al. 2001)	Indonesia	Pigs	208	6.4%	LAT
(Vostalová et al. 2000)	Czech Republic	Pigs	787	0.5%	RFC
(Arko-Mensah et al. 2000)	Ghana	Pigs	------	39%	ELISA

Table 6. *Cont.*

(Author Name/Year)	Location	Species	Examined Samples (No.)	Positive Samples (%)	Methods
(Suaréz-Aranda et al. 2000)	Brazil/Sao Paulo	Pigs	300	9.6%	ELISA
	Peru/Lima		96	32.3%	
(Venturini et al. 1999)	Argentine	Pigs (abortions)	738	2.03%	IFAT
				1.35%	MAT
(Weigel et al. 1999)	USA/Illinois	Pigs	174	31%	MAT
(Gamble et al. 1999)	New England	Pigs	1897	47.4%	MAT
(Gajadhar et al. 1998)	Canada	Pigs	2800	8.57%	LAT
(Davies et al. 1998)	USA/North Carolina	Pigs (fattening)	2238	0.58%	ELISA
(Dubey et al. 1997)	Georgia/Ossabaw Island	Pigs	1264	0.9%	MAT
(Dubey et al. 1995a)	USA/Illinois	Pigs (fattening)	4252	2.3%	MAT
(Dubey et al. 1995b)	USA/Iowa	Sows	2617	15.1%	MAT
(Weigel et al. 1995)	USA/Illinois	Sows	30	19.5%	Serology
(van Knapen et al. 1995)	Netherlands	Sows	-------	30.9%	Serology
(Edelhofer 1994)	Austria	Pigs	4697	13.7%	IFAT
(Dubey et al. 1991)	USA	Pigs	11842	23.9%	MAT
(Zimmerman et al. 1990)	USA/Iowa	Pigs (fattening)	2100	5.4%	ELISA
		Sows	616	11.4%	
WILD BOARS (*Sus scrofa*)					
(Machado et al. 2021)	Brazil/São Paulo	Wild boars	26	76.9%	ELISA
(Bier et al. 2020)	Germany	Wild boars—abdominal or thoracic fluid	180	24.4%	ELISA

Table 6. Cont.

(Author Name/Year)	Location	Species	Examined Samples (No.)	Positive Samples (%)	Methods
(Laforet et al. 2019)	Denmark	Wild boars	101	29.7%	ELISA
(Malmsten et al. 2018)	Sweden	Wild boars	276	29%	ELISA
(Roqueplo et al. 2017)	France/Var	Wild boars	841	16.8%	MAT
(Waap et al. 2016)	Portugal	Wild boar	26	7.7%	DAT
(Wallander et al. 2015)	Sweden	Wild boars	1327	49.5%	ELISA
(Coelho et al. 2014)	Portugal	Wild boars	97	20.6%	MAT
(Deksne and Kirjušina 2013)	Latvia	Wild boars (meat juice)	606	33.3%	ELISA
(Jokelainen et al. 2012a)	Finland	Wild boars	197	33%	DAT
(Hotea et al. 2011a)	Romania/ Timis County	Wild boars	52	94.23%	ELISA
(Berger-Schoch et al. 2011)	Switzerland	Wild boars	150	0.7%	PCR
(Richomme et al. 2009)	France	Wild boars	148	17.6%	MAT
(Fornazari et al. 2009)	Brazil/Sao Paulo	Wild boars	306	4.5%	MAT
(Antolová et al. 2007)	Slovak Republic	Wild boars	320	8.1%	ELISA
(Omata et al. 2005a)	Japan/ Shikoku	Wild boars	115	0%	ELISA
(Bártová et al. 2006)	Czech Republic	Wild boars	565	26.2%	IFAT
(Gauss et al. 2005)	Spain	Wild boars	507	38.4%	MAT
(Nogami et al. 1999)	Japan/ Iriomote	Wild boars	108	5.6%	LAT

Source: Table by authors.

Garcia et al. (2008) tested four serological methods and PCR for the diagnosis of ocular toxoplasmosis in pigs. Eighteen pigs of different breeds were experimentally infested with 4×10^4 *T. gondii* oocysts and were euthanized at 60 days. Serum, aqueous humor and retinal samples were collected. Serum and aqueous humor were tested by IFAT, MAT, ELISA, IB and retinal samples by PCR. Of all four serological methods, Ig G antibodies were found to be higher in serum than in aqueous humor. Through PCR, the parasite could be isolated from three samples. This suggests that serological methods alone, or PCR as the only method for determining ocular toxoplasmosis, are not sufficient. These methods must be corroborated to obtain a real result (Garcia et al. 2006a, 2006b).

Kijlstra et al. (2008) draw attention to the role of rodents in transmitting *T. gondii* infection to pigs. They tested various rodent species from three farms in the Netherlands via PCR and identified a prevalence of 10.3% in *Rattus norvegicus*, 6.5% in *Mus musculus*, 14.3% in *Apodemus sylvaticus* and 13.6% *Cricidura russula*. Within the three farms, a rodent control campaign was applied for a period of 4 months. The prevalence of *T. gondii* infection in pigs varied between 8 and 17%, decreasing during the 4 months to values between 0 and 10%. After the end of the rodent control campaign, in two farms, the prevalence was 0%, and in the third farm, the infection started to reappear after a while. This study proves the importance of rodents in transmitting *T. gondii* infection to humans and animals, acting as a reservoir for this parasite (Kijlstra et al. 2008).

Srikitjakarn et al. (2008), at the International Congress of Infectious Diseases, showed a decrease in the prevalence of *T. gondii* infection in pigs over the years. Serum samples from small, medium and large (industrial) farms in northern Thailand were tested. The LAT showed a prevalence of 7%, less than five times the prevalence shown 15 years ago. The ELISA showed a prevalence of only 0.2% (Srikitjakarn et al. 2008).

The LAT method was also tested by Jiang et al. (2008) in China, highlighting the specificity, stability and reproducibility of the method. Jiang pointed out that LAT is suitable for the early determination of anti-*Toxoplasma* antibodies (Jiang et al. 2008).

In addition to the data described, the prevalence of *T. gondii* infection in pigs has been studied by many well-known authors in the field in different countries (Table 6).

In order to obtain a low prevalence of *T. gondii* infection in pig farms, farm management, especially hygiene conditions, should be corrected with the introduction of mechanized cleaning, as well as the control of rodents and other species of mammals and birds on farms (Dubey 2009c; Dubey et al. 2014a). Rigorous and permanent control of rodents on animal farms for human consumption could be a solution to prevent toxoplasmosis in humans. As far as consumers are concerned, they should be informed of the risk of transmitting *T. gondii* infection through the handling of raw meat or the consumption of insufficiently heat-treated meat or meat preparations.

Domestic Birds

Chickens

Chickens (*Gallus domesticus*) raised on the ground are a potential source of infestation due to their increased level of exposure to environmental oocysts. All three genotypes of *T. gondii* could be isolated from the tissues of naturally infected chickens. Usually, these birds do not develop clinical signs of the disease, even after oral administration with a large number of oocysts. Egg production can be adversely affected in eggs infested with a large number of oocysts, but *T. gondii* is not transmitted through eggs. Toxoplasmosis does not occur on new farms, where chickens are raised in the barn without access to the outside. In a recent study by Dubey (2005) (cited by Weiss and Kim 2004), 2094 poultry samples were tested in major supermarkets in the United States, and none could be identified in the parasite (Weiss and Kim 2004).

Most studies on *Toxoplasma gondii* infection in chickens have focused on establishing seroprevalence as well as genetic characterization of isolates (Table 7). This is due to the fact that a very high genetic diversity of *T. gondii* has been observed in chickens.

The author of most studies on chickens is Dubey. In 2006, Dubey et al. established the prevalence of *T. gondii* infection in chickens in Nicaragua, Central America. The authors tested 98 chickens (by MAT) and obtained a prevalence of 85.7%. Tissue samples (heart and brain) were inoculated into seronegative cats and mice, yielding 48 isolates. They were classified into eight genotypes, especially types I, II, III or a combination thereof. It was observed that several genotypes were isolated from chickens in the same yard, suggesting that multiple genotypes may "circulate" in the same environment. This explains the increased frequency with which mixed infections (with several genotypes) were identified on the same individuals (Dubey et al. 2005b, 2006f).

Table 7. Seroprevalence of *T. gondii* infection in domestic birds.

Author Name (Year)	Location	Species	Examinated Samples (No.)	Positive Samples (%)	Methods
(Chaklu et al. 2020)	Ethiopia	Backyard chickens	384	72.4%	LAT
(Rodrigues et al. 2019)	Portugal	Free-range chickens	178	5.6%	MAT
		Indoor broiler chickens	170	0	
(Pardini et al. 2016)	Argentina/ Misiones	Free range chickens – brain tissue	33	30.3%	Nested-PCR
(Beltrame et al. 2012)	Brazil/ Espírito Santo	Free-range chickens	510	40.4%	IHAT
				38.8%	MAT
(Dubey et al. 2010a)	Brazil/ Fernando de Noronha	Hens (free bred)	50	84%	MAT
(Aigner et al. 2010)	Brazil	Chickens	65	43.07%	MAT

Table 7. *Cont.*

Author Name (Year)	Location	Species	Examinated Samples (No.)	Positive Samples (%)	Methods
(Harfoush and Tahoon 2010)	Egypt	Hens (free bred)	-------	38.1%	IHAT
		Ducks		55%	
		Turkeys		29.4%	
(Bártová et al. 2009b)	Czech Republic	Chickens	217	0%	IFAT
		Chickens (broilers)	293	0.34%	
		Ducks	360	14%	
		Geese	178	43%	
		Turkeys	60	0%	
(Yan et al. 2009)	Guangdong	Hens: (free bred)	361	11.4%	MAT
		Hens: (raised in battery cages)	244	4.1%	
		Ducks	349	16%	
(Zhu et al. 2008)	China	Hens: (free bred)	308	34.7%	ELISA
		Hens: (raised in battery cages)	210	2.8%	
(Dubey et al. 2007e)	Brazil/Rio Grande do Sul	Hens	84	46.4%	MAT
(Dubey et al. 2006e)	Portugal	Hens	225	27.11	MAT
(Dubey et al. 2006g)	Brazil/ Amazon	Hens	50	66%	MAT
(Dubey et al. 2006b)	USA/Costa Rica	Hens	114	40.1%	MAT
(Brandão et al. 2006)	Brazil/Minas Gerais	Hens (from households)	28	53.6%	Serology
		Hens (intensive system)	50	0%	

Table 7. Cont.

Author Name (Year)	Location	Species	Examined Samples (No.)	Positive Samples (%)	Methods
(Dubey et al. 2005a)	Peru	Hens	50	20%	MAT
(Dubey et al. 2005b)	Austria	Hens	830	36.3%	MAT
(Dubey et al. 2005c)	India/Grenada	Hens	102	52%	MAT
(Dubey et al. 2004b)	Mexico	Hens	208	6.2%	MAT
(Sreekumar et al. 2003)	India	Hens	741	17.9%	MAT
(Dubey et al. 2003b)	Brazil/Parana	Hens	40	40%	MAT

Source: Table by authors.

The same author, Dubey (Dubey et al. 2007a) conducted a study on poultry in a single yard in Illinois, USA. He identified three chickens with neurological signs (torticollis, inability to maintain balance, walking sideways). One of the hens was examined post mortem and the presence of non-suppurative encephalitis was observed, with numerous *T. gondii* tachyzoites, free and trapped in the tissue. The other 11 chickens and 1 goose from the same yard were examined serologically (by MAT) to identify *T. gondii* infection. All birds developed a positive response. Tissue inoculations (heart, brain and skeletal muscle) were performed on mice and viable forms of *T. gondii* were identified. All isolates belonged to genotype II (Dubey et al. 2007d).

In 2008, Dubey et al. examined 151 chickens from Brazil and obtained 58 genotypes. The results confirm the genetic diversity of *T. gondii* isolates in Brazil (Dubey et al. 2002c, 2008d).

The role of birds in transmitting the infection to humans, through eggs or infested meat, is not fully clarified, requiring further study.

Turkeys

Domestic turkeys (*Meleagris gallopavo*) infested with oocysts of *T. gondii* do not show clinical forms. However, some may have Aspergillus-associated pneumonia, and the parasite may have been isolated from their chest and thighs. Clinical toxoplasmosis does not occur in modern turkey-intensive farms (Weiss and Kim 2004).

Ducks and geese

Domestic ducks (*Anas platyrhynchos*) infested with *T. gondii* oocysts show no clinical signs, but genotype II of the parasite could be isolated from some naturally infected ducks. Regarding goose infection, there are no data on the isolation of parasitic forms or on trials of experimental infection (Weiss and Kim 2004).

Pigeons

These birds coexist with humans in both rural and urban areas. Pigeon meat, eaten by humans or other carnivorous species, can be a source of *Toxoplasma gondii* infection. de Lima et al. (2011) tested 238 samples from pigeons in Sao Paolo, Brazil. Serological tests (MAT) showed a prevalence of 5%. These data confirm the presence of the parasite in this species, but with a low prevalence (de Lima et al. 2011).

Waap et al. (2008) genetically characterized *T. gondii* isolates obtained from pigeons in Lisbon, Portugal. Six-hundred and ninety-five pigeons from the city were examined by DAT, and a prevalence of 4.6% was obtained. Twelve isolates belonging to types I, II and III were identified. These data indicate the worldwide distribution of this parasite. Birds are considered a good indicator for soil infection with *T. gondii*, due to their tendency to feed directly from the soil (Waap et al. 2008).

Toxoplasmosis in pigeons has been reported by other authors (Table 8).

Table 8. Seroprevalence of *T. gondii* infection in pigeons.

(Author Name/Year)	Location	Species	Examined Samples (No.)	Positive Samples (%)	Methods
(Cano-Terriza et al. 2015)	Spain	Feral pigeon	142	9.2%	ELISA
(Waap et al. 2012)	Portugal/ Lisbon	Pigeons	1507	2.6%	DAT
(Yan et al. 2011)	China/ Guangdong	Pigeons	275	18.2%	MAT
(Salant et al. 2009a)	Israel	Pigeons	495	4%	MAT
(Piasecki et al. 2004)	Poland	Pigeons	230	74.8%	LAT

Source: Table by authors.

Horses

Anti-*Toxoplasma* antibodies have been identified in horse serum by several authors (Table 9). In 2007, Tassi reviewed details about the pathology and epidemiology of toxoplasmosis in horses, aspects previously described by other authors (Tassi 2007). Clinical forms of toxoplasmosis have not been described in horses, but some researchers have attempted to identify the degree of involvement of *T. gondii* in some cases of encephalomyelitis (Dubey et al. 1999a; 2001; Turner and Savva 1991).

Table 9. Seroprevalence of T. gondii infection in horses.

(Author Name/Year)	Location	Species	Examined Samples (No.)	Positive Samples (%)	Methods
(Rissanen et al. 2019)	Ukraine/ Kyiv, Lviv	Horses	76	7.9%	ELISA
(Munhoz et al. 2019)	Brazil/Bahia	Horses (Farms)	475	27.3%	IFA
		Donkeys	33	72.7%	
(Ouslimani et al. 2019)	Algeria	Horses (Farms)	736	18.1%	LAT
(Schale et al. 2018)	USA/Eastern	Horses	101	14.9%	Western Blot
(Xing et al. 2018)	China/Xinjiang	Horses (Farms)	409	5.4%	ELISA
(Guerra et al. 2018)	Brazil/ Pernambuco	Horses (Farms)	387	12.4%	MAT
(James et al. 2017)	USA/ California	Horses (Clinics)	392	62%	IFA
(Klun et al. 2017)	Serbia/ Northern	Horses (Abattoirs)	105	48.6%	MAT
(Bártová et al. 2017b)	Nigeria	Horses (Farms)	120	24%	IFA
		Donkeys	24	17%	
(Masatani et al. 2016)	Japan	Horses	783	5.1%	LAT
(Razmi et al. 2016)	Iran/ Khorasan	Horses	100	14%	IFA
(Cazarotto et al. 2016)	Brazil/Santa Catarina	Horses (Farms)	174	32.2%	IFA
(Paştiu et al. 2015a)	Romania	Horses (Abattoirs)	82	37.8%	MAT
				39%	ELISA
(Papini et al. 2015)	Italy/La Spezia	Horses (Abattoirs)	153	17.6%	IFA
(Aroussi et al. 2015)	France	Horses	136	58.4%	ELISA
(Mancianti et al. 2014)	Italy/Tuscany	Horses (Farms)	44	25%	IFA
(Ayinmode et al. 2014)	Nigeria/ Southwestern	Horses	157	14%	MAT

Table 9. *Cont.*

(Author Name/Year)	Location	Species	Examined Samples (No.)	Positive Samples (%)	Methods
(Lee et al. 2014)	Korea	Horses (Farms)	816	2.9%	ELISA
(Dubey et al. 2014b)	USA	Donkeys	373	6.4%	MAT
(Lopes et al. 2013)	Portugal/North	Horses	173	13.3%	MAT
(Asgari et al. 2013)	Iran/Fars	Horses (Farms)	35	40%	MAT
(Finger et al. 2013)	Brazil/Paraná	Carthorses	100	17%	IFA
(García-Bocanegra et al. 2012)	Spain/Andalusia	Horses	454	10.8%	MAT
		Donkeys	82	25.6%	
		Mules	80	15%	
(Alvarado-Esquivel et al. 2012c)	Mexico/Durango	Domestic Horses	495	6.1%	MAT
(Costa et al. 2012)	Brazil/Fernando de Noronha	Horses	16	43.7%	MAT
(Hotea et al. 2011a)	Romania/Caras-Severin County	Horses	650	32.77%	ELISA
(Hotea et al. 2011a)	Romania/Arad County	Horses	950	29.79%	ELISA
(Hotea et al. 2011a)	Romania/Timis County	Horses	1400	37.92%	ELISA
(Kouam et al. 2010)	Greece	Horses	753	1.7%	ELISA
		Mules	13	7.7%	
		Ponies	7	0%	
(Karatepe et al. 2010)	Turkey/Nigde	Horses	125	7.2%	SFDT
(Haridy et al. 2010)	Egypt/Cairo	Donkeys	100	45%	ELISA
(Goez et al. 2007)	Turkey/Hakkari	Horses	74	28.4%	IHA
(Güçlü et al. 2007)	Turkey/Ankara	Horses	100	28%	SFDT

Table 9. *Cont.*

(Author Name/Year)	Location	Species	Examined Samples (No.)	Positive Samples (%)	Methods
(Naves et al. 2005)	Brazil	Horses	117	12.82%	IFAT
(Jakubek et al. 2006)	Sweden	Horses	414	1%	MAT
(Silva 2005)	Brazil/ Pantanal	Horses	150	1.33%	IHA
(Sevgili et al. 2004)	Turkey/ Sanhurfa	Horses	93	7.5%	SFDT
(Akca et al. 2023)	Turkey/Kars	Horses	189	20.6%	SFDT
(Dubey et al. 2003a)	USA/ Wyoming	Wild Horses	276	0.36%	MAT
(Gupta et al. 2002)	South Korea/Jeju	Horses	191	2.6%	IFAT
(Aslantaş et al. 2001)	Turkey/Kars	Horses	111	1.8%	SFDT
(Wyss et al. 2000)	Switzerland	Horses	120	4%	ELISA
(Dubey et al. 1999a)	North America	Horses	1788	6.9%	MAT
(Dubey et al. 1999b)	Brazil	Horses	101	15.84%	MAT
(Garcia et al. 1999)	Brazil/Parana	Horses	173	12.1%	IFAT
(Dubey et al. 1999c)	Argentine	Horses	76	13.1%	MAT
(El-Ghaysh 1998)	Egypt/Monofia	Donkeys	121	65.6%	ELISA
(Gazeta et al. 1997)	Brazil/Rio de Janeiro	Horses	430	4.4%	IFAT
(Inci and Karaer 1996)	Turkey/Gemlik	Horses	103	2%	SFDT
(Hejlicek and Literak 1994b)	Czech Republic	Horses	2886	7.7%	SFDT
(Uggla et al. 1990)	Sweden	Horses	219	1%	ELISA

Source: Table by authors.

In England, Turner and Savva (1990, 1992) identified *T. gondii* DNA in the placenta of an aborted mare, and later, Marques et al. (1995) attempted to demonstrate transplacental transmission to this species (Turner and Savva 1990, 1992; Marques et al. 1995). He experimentally infested nine pregnant mares with *T. gondii* oocysts, and then was able to isolate the parasite from the tissues of the foals.

Horses appear to be resistant to experimental infection with 1×10^4 or 1×10^5 oocysts (Weiss and Kim 2004), but administration of 20 million and 32 million tachyzoites of the RH strain to two horses, respectively, resulted in fever within the first 4–8 days (Ghazy et al. 2007). On days 6–12 p.m. the animals showed an anti-*Toxoplasma* immune response, but on day 30 p.i., antibodies could no longer be detected (Sposito Filha et al. 1992).

Güçlü et al. (2007) tested 100 clinically healthy horses from sports breeds. The horses came from Ankara, Turkey. Serum samples were tested by the SFDT technique and a seroprevalence of 28% was obtained, with a positivity of 30.7% in males and 20% in females. It is believed that the animals became infected as a result of contact with soil contaminated with *T. gondii* oocysts (Figure 10). The high prevalence obtained indicates the need for further studies regarding the ability of the parasite to form cysts in the meat of horses destined for slaughter (Güçlü et al. 2007).

Figure 10. Breeding conditions for horses. Source: Photos by authors.

The increased prevalence obtained by these studies draw attention to the possibility of transmitting *T. gondii* infection to humans or other animals through the consumption of insufficiently heat-treated meat or meat preparations (Table 9).

Cattle

Clinical toxoplasmosis in cattle is rare, and *Toxoplasma*-induced abortion is unusual. Many of the data on abortion are currently considered to be due to infection with *Neospora caninum*. The intention to isolate the parasite from HIV-positive cattle has failed, pointing out that beef preparations are not a very important source of infection for humans. In the United States, 2094 beef samples were withdrawn from supermarkets and an attempt was made to isolate the parasite, but without positive results. However, tissue cysts may remain viable in cattle for more than 1191 days. In these cases, additional studies appear to be needed (Weiss and Kim 2004).

Moré et al. (2008a, 2008b) established the prevalence of *Sarcocystis cruzi*, *Neospora caninum* and *Toxoplasma gondii* infections in cattle in Argentina. Ninety beef cows were examined by IFAT and the following prevalence results were obtained: all cows responded positively for S. cruzi, 73% for *N. caninum* and 91% for *T. gondii*. Diaphragm, esophagus and myocardial samples from seropositive animals were examined directly under a microscope, in histological preparations, by immuno-histochemistry (IHC) and PCR. *S. cruzi* was identified in all myocardial samples. *T. gondii* and *N. caninum* were not identified by IHC, and *Toxoplasma gondii* DNA was identified in two myocardial samples from the 20 HIV-positive animals. *N. caninum* DNA could not be identified. Infections associated with the three parasites were observed in 70% of the examined cows, infections with *T. gondii* and *S. cruzi* in 21% and *N. caninum* and *S. cruzi* in 34% (Moré et al. 2008a, 2008b).

Moore et al. (2008) highlighted the role of *Neospora caninum* and *T. gondii* infections in spontaneous abortions of cattle in Argentina. Based on the histopathological results, it was considered that, in 70 of the cows, the cause of the abortion was parasitic, out of a total of 666 cattle that suffered spontaneous abortions. Nerve tissue samples (IFAT, IHC or PCR) from aborted fetuses were tested for *N. caninum* infection and 66 (9.9%) tested positive. *T. gondii* infection could not be diagnosed in any of the abortions. This study highlights the hypothesis that *N. caninum* infection is one of the common causes of miscarriages in cattle, but not *T. gondii* infection (Moore et al. 2008).

In addition to the data described above, the prevalence of *T. gondii* infection in catlle has also been studied by other well-known authors in the field in other countries (Table 10).

Table 10. Seroprevalence of *T. gondii* infection in cattle.

(Author Name/Year)	Location	Species	Examinated Samples (No.)	Positive Samples (%)	Methods
(Sah et al. 2018)	Bangladesh	Cattle	252	8.33%	ELISA
(Jokelainen et al. 2017)	Estonia	Cattle	3991	18.62%	DAT
(Allén 2016)	Finland	Cattle (meat juice)	200	7.5%	ELISA
(Lopes et al. 2013)	Portugal/ North	Cattle	161	7.5%	MAT
(Hotea et al. 2011a)	Romania/ Caras-Severin County	Cattle	1100	17.9%	ELISA
(Hotea et al. 2011a)	Romania/ Arad County	Cattle	850	14.23%	ELISA
(Hotea et al. 2011a)	Romania/ Timis County	Cattle	1500	25.4%	ELISA

(Author Name/Year)	Location	Species	Examined Samples (No.)	Positive Samples (%)	Methods
(Berger-Schoch et al. 2011)	Switzerland	Cattle	406	4.7%	PCR
(Costa et al. 2011)	Brazil/Sao Paulo	Cattle (pregnant)	50	18%	IFAT
(Opsteegh et al. 2011)	Netherlands	Cattle (farm 1)	638	22.7%	ELISA
		Cattle (farm 2)	268	54.5%	
(Panadero et al. 2010)	Spain/Galicia	Cattle	178	7.3%	DAT
(Santos et al. 2010)	Brazil/Bahia	Cattle (meat)	100	26%	IFAT
(Santos et al. 2009)	Brazil/Mato Grosso	Cattle (dairy)	2000	71%	IFAT
(Gilot-Fromont et al. 2009)	France/ Champagne-Ardenne	Cattle	1329	7.8%	MAT
(Yildiz et al. 2009)	Turkey	Cattle (aborted)	234	15.81%	SFDT
		Cattle (pregnant)	323	31.36%	
(Yu et al. 2007)	China	Cattle (dairy)	262	2.3%	IAT
		Cattle (meat)	10	0%	
(Albuquerque et al. 2005)	Brazil/Rio de Janeiro	Cattle	589	65%	IFAT
(Klun et al. 2006)	Serbia	Cattle	611	76.3%	MAT
(Meireles et al. 2003)	Brazil/Sao Paulo	Cattle	200	11%	ELISA
(Karatepe et al. 2004)	Turkey/ Nigde	Cattle	100	47%	SFDT
(Ortega et al. 2003)	USA	Cattle	76	43%	IB
	Peru		253	50%	
(van Knapen et al. 1995)	Netherlands	Cattle	189	27.9%	Serology

Following these studies, attention is drawn to existing co-infestations in cattle, infections that cause a decrease in the number of products (in neosporosis), confiscation of meat in slaughterhouses (in sarcocystosis) or the possibility of transmission to consumers (Figure 11).

Figure 11. Cats access in cattle shelters. Source: Photos by authors.

Water buffaloes

The natural infection with the expression of the clinical form of the disease could not be identified in buffaloes (*Bison bison, Bubalus bubalis, Syncerus caffer*) and the parasite could not be isolated from these species (Weiss and Kim 2004). Yu et al. (2007) tested 40 buffaloes in China and did not identify any positive evidence (Yu et al. 2007). Additionally, Hotea et al. (2011a) analyzed serological samples from 92 buffaloes, from Romania, using the ELISA technique, and none of the investigated samples showed anti-*Toxoplasma* antibodies (Figure 12) (Hotea et al. 2011b). Although these species are considered resistant to *T. gondii* infection, Gencay et al. 2013 identified tissue cysts in buffalo meat in 3 of 20 examined samples (15%) (Gencay et al. 2013).

Figure 12. Water buffaloes raised in Romania. Source: Photos by authors.

Camels

An episode of acute toxoplasmosis was identified in a 6-year-old camel (*Camelus dromedarius*). The dominant symptom was dyspnea, and tachyzoites could be identified in the lungs and exudates. *Toxoplasma gondii* could also be isolated from camel meat (Weiss and Kim 2004).

Llama, alpaca and vicuña

Experimental studies indicate that llama (*Lama glama*) is resistant to *T. gondii* infection, even if inoculation was performed during pregnancy. Natural infections with *T. gondii* have not been observed in llama, alpaca (*Lama pacos*) or vicuna (*Lama vicugna*) (Weiss and Kim 2004).

In the world, the prevalence of *T. gondii* infection in these species is varied (Table 11).

Table 11. Seroprevalence of *T. gondii* infection in llama, alpaca and vicuña.

(Author Name/Year)	Location	Species	Examined Samples (No.)	Positive Samples (%)	Methods
(Moré et al. 2008a, 2008b)	Argentine	Llama (*Lama glama*)	308	30%	IFAT
(Wolf et al. 2005)	Peru	Alpaca (*Lama pacos*), Llama (*Lama glama*) and Vicuna (*Lama vicugna*)	871	3.44%	IFAT
	Germany		32	43.75%	
(Chávez-Velásquez et al. 2005)	Peru	Llama (*Lama glama*)	43	55.8%	IFAT
		Vicuna (*Vicugna vicugna*)	200	5.5%	
(Gorman et al. 1999)	Chile	Alpaca (*Lama pacos*)	447	16.3%	IFAT

Source: Table by authors.

Minks

Acute toxoplasmosis with abortion has been found in farm mink (*Mustela vison*) in Europe and the United States. At one of the farms, the consumption of fresh, unfrozen slaughterhouse organs was incriminated. *Toxoplasma gondii* has also been isolated from wild minks in the United States (Weiss and Kim 2004).

An episode of acute toxoplasmosis was reported by Frank, 2001 on a Wisconsin farm, when 26% of the 7800 minks miscarried or gave birth to dead chicks. Cytomegalovirus infection could not be identified, which had happened a year earlier. *T. gondii* infection was confirmed histopathological, with lesions in the brain, heart, liver and lungs. Tachyzoites were identified in these lesions, and final confirmation was made via immunohistochemistry (Frank 2001).

Toxoplasmosis in minks has been reported in other countries (Table 12).

Table 12. Seroprevalence of T. gondii infection in minks.

(Author Name/Year)	Location	Species	Examined Samples (No.)	Positive Samples (%)	Methods
(Sepúlveda et al. 2011)	Chile	Minks	30	70%	LAT
(Smielewska-Łoś and Turniak 2004)	Poland	Minks	961	13.9%	LAT
(Arnaudov 2003)	Bulgaria	Minks	156	12.2%	ELISA
(Henriksen et al. 1994)	Denmark	Minks	195	3%	LAT

Source: Table by authors.

4.6.2. T. gondii Infection in Pets

Ferrets

Congenital toxoplasmosis has also been observed in domestic ferrets (*Mustela putorius furo*) in New Zealand and 13% of chickens have died from the acute form of the disease (Weiss and Kim 2004). An epizootic episode of toxoplasmosis was found in a ferret population (*Mustela nigripes*) in zoos in Kentucky, United States, when 22 adults and 30 chicks died of acute toxoplasmosis and 13 adults also died due to the chronic form of the disease. Nineteen adults and six pups showed symptoms such as anorexia, corneal edema and ataxia. Paraclinical (by MAT and LAT), very high titers of anti-*Toxoplasma* antibodies were observed, and the confirmation of the infection was made by immunohistochemical technique and ultrastructural examination (Burns et al. 2003). A similar case of toxoplasmosis was reported in New Zealand (Dubey 2010) when 270 ferrets out of 750, died within a few weeks with no apparent clinical signs. Myocarditis and hepatitis were the main lesions identified.

Cats

All domestic cats, of any age, sex or breed, are susceptible to *T. gondii* infection. Transplacental or lactogenic infected kittens can eliminate oocysts, but the prepatent period is 3 weeks or more due to the fact that the infection occurs with tachyzoites. Domestic cats under the age of one eliminate a very large number of oocysts. Cats born or raised outside their homes are infected with *T. gondii* immediately after they are weaned and begin to hunt.

Cell-mediated intestinal immunity to *T. gondii* is very strong in cats that have eliminated oocysts. The first infection in cats does not cause immunosuppression. Antibodies in the blood do not play a very important role in resistance to intestinal infection. Oocysts begin to be excreted in the feces before the appearance of Ig M, Ig G and Ig A immunoglobulins in the blood. Most cats that once eliminated oocysts will not re-eliminate oocysts on new contact with the parasite, with intestinal immunity lasting 6 years in 55% of the cats studied (Weiss and Kim 2004).

Recent research (Weiss and Kim 2004) has attempted to identify types of *T. gondii* isolated from naturally infected cats. Dubey et al. (2004c) were able to isolate 37 types of *T. gondii* from 54 cats in Paraná, Brazil (Dubey et al. 2004c). Of the 37 isolated types, 15 were genotype I and 22 were genotype III. In these cases, the parasite was more frequently isolated from the heart than from the brain. Genotype II was isolated from two of two cats tested in Mississippi. Pena (2006) (cited by Weiss and Kim 2004) isolated the parasite from the tissues of 47 of 71 cats in Sao Paolo, Brazil. Of these, 34 were genotype I, 12 genotype III, and a mixed one, genotypes I and III (Weiss and Kim 2004).

Theoretically, cat owners and veterinarians are believed to be most at risk of developing toxoplasmosis, but serological studies do not confirm this claim. Research shows, however, that cat owners are less at risk, but studies continue. Attempts have also been made to make correlations between people who own cats or those who come into contact with cats and people who are positive for *Toxoplasma* infection. Some correlations came out positive, others negative. This emphasizes that taking measures regarding contact with cats is not necessarily a prevention of *T. gondii* infection (Weiss and Kim 2004; Dubey et al. 2014a).

Pregnant women or immunocompromised people should not change the sand in cat litter. If the feces are discarded daily, the oocysts will increase. Oocysts last in the soil for years and can be spread, by chance, with various maneuvers performed on the ground or through vector hosts. In 1979, Teutsch (cited by Weiss and Kim 2004) described an episode of *Toxoplasma* infection in people who participated in a horse race (Weiss and Kim 2004). The oocysts are thought to have been inhaled from high dust on the runway. However, oocysts do not stay in the air for a long time, so inhalation infestation is rare. Washing fruits and vegetables, as well as wearing gloves during gardening, are measures to prevent exposure to *Toxoplasma gondii* oocysts (Figure 13).

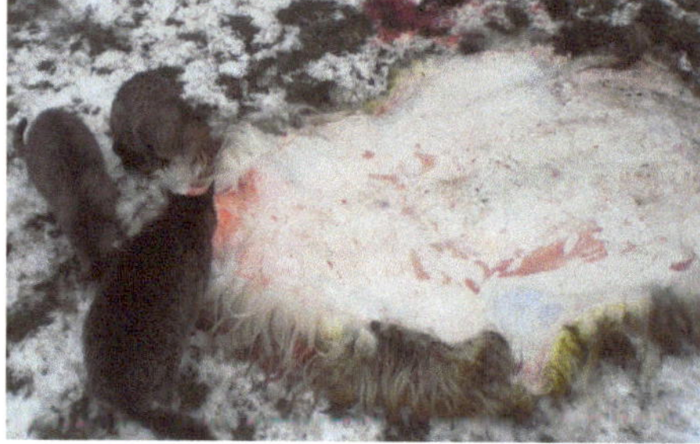

Figure 13. Cats freely living in households. Source: Photos by author.

Oocysts could not be isolated from the fur of oocyst-eliminating cats, so stroking cats is not a risk of infection. Additionally, tachyzoites are not present in the oral cavity of oocyst-eliminating cats, so cat bites are not a risk factor; nor do their scratches present danger of infestation (Weiss and Kim 2004).

Salant et al. (2007) developed a copro-diagnostic PCR method in toxoplasmosis. The study started from the premise that microscopic examination of feces is not sensitive enough to observe *T. gondii* oocysts. Additionally, inoculation on laboratory animals takes a long time and is expensive. Based on these considerations, the author adapted a PCR method to determine the DNA of fecal *T. gondii* oocysts. Of the 122 fecal samples obtained from cats in Jerusalem, 11 positive results were found, while microscopic examination did not show any positive samples (Salant et al. 2007).

It should be noted that cats begin the elimination of oocysts before antibodies form in the blood. This PCR method thus becomes very important in establishing an early diagnosis and the acute phase of elimination of *T. gondii* oocysts. Given that infestation of humans and animals is often achieved through contaminated water or soil, this method can also be adapted for the determination of *T. gondii* oocysts in water and soil samples.

In addition to the data described above, the serological prevalence of *T. gondii* infection in cats has also been studied by other well-known authors in the field from other countries (Table 13).

Table 13. Seroprevalence of *T. gondii* infection in cats.

(Author Name/Year)	Location	Species	Examined Samples (No.)	Positive Samples (%)	Methods
(Neves et al. 2020)	Portugal/ Madeira Island	Cats (pets)	141	30.5%	ELISA
(Qiu et al. 2020)	China/ Heilongjiang	Cats (pets)	352	19.6%	IHA
(Brennan et al. 2020)	Australia/ Sydney	Cats (pets)	417	39%	ELISA
(Yücesan et al. 2019)	Turkey/ Ankara	Cats (pets and stray)	129	66.6%	Dye test
(Ybañez et al. 2019)	Philippines/ Cebu	Cats (pets and stray)	104	42.3%	LAT
(Khodaverdi and Razmi 2019)	Iran/Mashhad	Cats (stray)	159	59.1%	ELISA
(Sroka et al. 2018)	Poland/ Southwestern	Cats (pets)	208	68.8%	IFAT
(Asgari et al. 2018)	Iran/Shiraz	Cats (stray)	145	82.8%	MAT
(Barros et al. 2018)	Chile/ Southern	Cats (pets)	65	67.6%	MAT
(Hou et al. 2018)	China/Jiangsu Province	Cats (stray)	64	34.38%	ELISA
(Montoya et al. 2018)	Spain/Central	Cats (stray)	356	24.2%	DAT
(Veronesi et al. 2017)	Italy/Perugia	Cats (pets)	78	42.3%	IFAT
(Must et al. 2017)	Finland	Cats	1121	41.1%	MAT

$$\textbf{Table 13. } \textit{Cont.}$$

(Author Name/Year)	Location	Species	Examinated Samples (No.)	Positive Samples (%)	Methods
(Yekkour et al. 2017)	Algeria/ Algiers	Cats (stray)	96	50%	MAT
(Lopes et al. 2017)	Angola/ Luanda	Domestic cats	102	3.9%	MAT
(Tavalla et al. 2017)	Iran	Cats (fecal samples)	486	7.2%	PCR
(Rengifo-Herrera et al. 2017)	Panama	Cats (pets)	120	25%	ELISA
(de Melo et al. 2016)	Brazil/ Noronha Island	Feral cats	31	58%	IFAT
(Verma et al. 2016)	USA/ Minnesota	Cats (pets)	16	56.2%	MAT
(Erkılıç et al. 2016)	Turkey/Kars	Cats (pets and stray)	102	44.1%	Dye test
(Spada et al. 2016)	Italy/Milan	Cats (stray)	82	29.3%	IFAT
(Yang et al. 2015)	China	Cats	42	50%	MAT
(Furtado et al. 2015)	Brazil	Cats (stray)	29	82.8%	IFAT
(de Souza et al. 2015)	Brazil/Acre	Cats (pets)	89	24.7%	IFAT
(Sævik et al. 2015)	Norway	Cats (pets)	478	41%	DAT
(Silaghi et al. 2014)	Albania/ Tirana	Cats (stray)	146	62.3%	IFAT
(Miró et al. 2014)	Spain/Madrid	Cats (stray)	346	53.4%	MAT
(Spada et al. 2013)	Italy/Milan	Cats (stray)	78	21.8%	IFAT
(Afonso et al. 2013)	France/ Central and Eastern	Cats (stray)	29	65.5%	MAT
(Jokelainen et al. 2012b)	Finland/ Helsinki	Cats (pets and from shelter)	490	48.4%	MAT

Table 13. Cont.

(Author Name/Year)	Location	Species	Examined Samples (No.)	Positive Samples (%)	Methods
(Qian et al. 2012)	China/Beijing	Cats (stray)	64	57.8%	MAT
(Györke et al. 2011)	Romania/ Cluj-Napoca	Cats (pets)	236	47%	ELISA
(Dărăbuş et al. 2011a)	Romania/ Western	Cats (stray)	173	66.4%	ELISA
(Hotea et al. 2011a)	Romania/ Caras-Severin County	Cats	62	77.42%	ELISA
(Hotea et al. 2011a)	Romania/ Arad County	Cats	78	80.77%	ELISA
(Hotea et al. 2011a)	Romania/ Timis County	Cats	265	62.26%	ELISA
(Hsu et al. 2011)	Virginia	Cats	232	27.15%	IFAT
(Berger-Schoch et al. 2011)	Switzerland	Cats (fecal samples)	252	0.4%	PCR
(Herrmann et al. 2010)	Germany/ Wuster-hausen	Cats	18.259	0.25%	PCR-RFLP
(Mircean et al. 2010)	Romania/Cluj Napoca	Cats	414	1.2%	Flotation method
(Akhtardanesh et al. 2010)	Iran	Cats	70	32.1%	ICA
(Montoya et al. 2009)	Spain/ Madrid	Cats	71	43.6% 33.8%	n-PCR rt-PCR
(Kamani et al. 2010)	Nigeria/ Maiduguri	Cats	105	36.2%	LAT
(Mancianti et al. 2010)	Italy/ Florence	Cats (serum samples)	50	44%	MAT
(Mancianti et al. 2010)	Italy/ Florence	Cats (fecal samples)	50	16%	n-PCR
(Montoya et al. 2009)	Spain	Cats	64	64%	Inoculation on mouse
(Montoya et al. 2009)	Spain	Cats	64	57.8%	PCR
(Millán et al. 2009)	Spain/ Balearic Islands	Cats	59	84.7%	MAT

(Author Name/Year)	Location	Species	Examined Samples (No.)	Positive Samples (%)	Methods
(Lee et al. 2008)	South Korea	Cats	106	47.2%	PCR
(Titilincu et al. 2008)	Romania/Cluj Napoca	Cats	50	81.08%	ELISA
(Antoniu et al. 2008)	Romania/Bucharest	Cats	42	47.61%	IFAT
				54.76%	ELISA
(Hornok et al. 2008)	Hungary/Budapest	Cats	330	47.6%	IFAT
(Yu et al. 2008)	China/Beijing	Cats	335	14.9%	ELISA
(De Craeye et al. 2008)	Belgium	Cats	567	25%	IFAT
(Besné-Mérida et al. 2008)	Mexico	Cats	169	21.8%	ELISA
(Lopes et al. 2008)	Portugal	Cats	204	35.8%	MAT
(Schares et al. 2008)	Germany + 16 other European countries	Cats (fecal samples)	24106	0.31%	Flotation method
(Bresciani et al. 2007)	Brazil/Sao Paulo	Cats	400	25%	IFAT
(Dubey et al. 2007g)	China	Cats	34	79.4%	MAT
(Salant et al. 2007)	Jerusalem	Cats (fecal samples)	122	9.01%	PCR
(Jittapalapong et al. 2007)	Thailand/Bangkok	Cats	592	11%	MLAT
(Javadi et al. 2008)	Iran/Uramia	Cats	30	30%	ELISA
(Chen et al. 2005)	China/Guangzhou	Cats	114	23.7%	ELISA
(Dubey 2006)	South America/Colombia	Cats	170	45.2%	MAT
(Asthana et al. 2006)	India/Granada	Cats	40	35%	MAT
(Cavalcante et al. 2006)	Brazil/Rondonia	Cats	63	87.3%	MAT and IFAT

Table 13. Cont.

(Author Name/Year)	Location	Species	Examined Samples (No.)	Positive Samples (%)	Methods
(Dubey et al. 2005c)	USA	Cats	58	84.4%	MAT
(Vollaire et al. 2005)	USA	Cats (clinic)	12628	31.6%	ELISA
(Nutter et al. 2004)	USA/North Carolina	Cats (stray) Cats (pets)	100 76	63% 34%	Serology
(Michalski and Platt-Samoraj 2004)	Poland	Cats	17	70.6%	DAT
(Salant et al. 2007)	Jerusalem	Cats	1062	16.8%	ELISA
(Miró et al. 2004)	Spain	Cats	585	32.3%	IFAT
(Porqueddu et al. 2004)	Italy/Sardinia	Cats	54	37%	IFAT
(Borkovcova 2003)	Czech Republic/Moravia	Cats (feces)	358	3.4%	Flotation method
(Gauss et al. 2003)	Spain/Barcelona	Cats	220	45%	MAT
(Schatzberg et al. 2003)	New York	Cats	7	100%	PCR
(Meireles et al. 2004)	Brazil/Sao Paulo	Cats	100	40%	ELISA
(Dubey et al. 2002d)	USA/Ohio	Cats	275	48%	MAT
(DeFeo et al. 2002)	Rhode Island	Cats	200	42%	MAT
(Silva et al. 2002)	Brazil/Sao Paulo	Cats	430	26.7%	MAT
(Maruyama et al. 1998)	Japan/Kanagaua and Saitama	Cats (pets)	471	8.7%	IFAT
(Dubey et al. 1995a)	USA/Illinois	Cats (farms)	391	68.3%	MAT

Source: Table by authors.

Dogs

In the past, toxoplasmosis in dogs was confused with neosporosis, so many of the data reported in the past as *Toxoplasma* infections are now considered infections with *Neospora caninum*. However, toxoplasmosis also develops in dogs. The clinical form is frequently associated with immunosuppression in the case of infection with the canine scabies virus, and the clinical signs concern, in particular, the respiratory and hepatic systems. Transplacental infection has not yet been identified in naturally infected dogs. Dogs are resistant to experimental infection with oocysts or tissue cysts (Weiss and Kim 2004; Dubey et al. 2006d; Dubey et al. 2014a).

The theory has been launched that these species play a major role in transmitting the infection to humans. Dogs that ingest oocysts can eliminate them in or around homes (Figure 14). It seems that in dogs' digestive tracts, oocysts are not destroyed by gastric enzymes. In experimental infections in dogs with oocysts, parasitic elements were found in the feces 2 days after infection. On the other hand, it seems that oocysts do not meet the necessary conditions for sporulation on the fur of dogs. So, this transmission path is not valid. Schares et al. (2005) identified viable parasitic forms in 2 of the 24,089 dogs examined in Germany (Schares et al. 2008). The role of dogs as possible vector hosts in the transmission of toxoplasmosis requires further research (Weiss and Kim 2004; Dubey et al. 2007b).

Figure 14. Raising conditions for dogs. Source: Photos by authors.

The prevalence of *T. gondii* infection in dogs has been studied by many well-known authors in the field in other countries (Table 14).

Table 14. Seroprevalence of *T. gondii* infection in dogs.

(Author Name/Year)	Location	Species	Examined Samples (No.)	Positive Samples (%)	Methods
(Fábrega et al. 2020)	Panama	Dogs	319	25.7%	ELISA
(da Cunha et al. 2020)	Brazil/Parana	Dogs (pets)	264	7.9%	IFA
(Coelho et al. 2019)	United Kingdom	Dogs (neurological)	201	3.9%	IFA

Table 14. *Cont.*

(Author Name/Year)	Location	Species	Examined Samples (No.)	Positive Samples (%)	Methods
(de Oliveira et al. 2019)	Brazil/Bahia	Dogs (pets)	61	44.3%	IFA
(Enriquez et al. 2019)	Argentina/Chaco	Dogs (pets)	152	55.3%	IFA
(Cong et al. 2018)	China/Shandong	Dogs (pets)	180	15.6%	IHA
(Fernandes et al. 2018)	Brazil/Paraíba	Dogs (pets)	1043	22.1%	IFA
(Zarra-Nezhad et al. 2017)	Iran/Ahvaz	Dogs (pets)	180	46.6%	ELISA
(Navarrete et al. 2017)	Cuba	Dogs	176	72.7%	ELISA
(Magalhães et al. 2017)	Brazil/Fernando de Noronha	Dogs (pets)	320	48.7%	IFA
(Verma et al. 2016)	USA/Minnesota	Dogs (pets)	14	42.8%	MAT
(Cano-Terriza et al. 2016)	Spain/Andalusia and Ceuta	Dogs (pets, clinics)	769	30.6%	MAT
(Machacova et al. 2016)	Italy/Avellino and Salerno	Dogs (for hunting)	398	24%	IFA
(Davoust et al. 2015)	Senegal	Dogs	145	67%	MAT
(Li et al. 2015)	China/Guizhou	Dogs (pets)	107	20.6%	ELISA
(de Seabra et al. 2015)	Brazil/Sao Paulo	Dogs (clinic)	300	24.3%	IFA
(Ahmad et al. 2014)	Pakistan/Potohar	Dogs (pets)	408	28.4%	ELISA
(Alvarado-Esquivel et al. 2014)	Mexico/Veracruz	Dogs (shelter)	101	67.3%	MAT
(Deng et al. 2011)	China/Changsha	Dogs (pets)	429	23.7%	ELISA
(Muz et al. 2013)	Turkey/Mediterranean Hatay	Dogs (Shepherd)	46	58.7%	ELISA

Table 14. *Cont.*

(Author Name/Year)	Location	Species	Examined Samples (No.)	Positive Samples (%)	Methods
(Arunvipas et al. 2013)	Thailand	Dogs (farms)	114	6.1%	LAT
(El Behairy et al. 2013)	Egypt/Giza	Dogs (stray)	51	98%	MAT
(Sroka and Szymańska 2012)	Poland/Lublin	Dogs (pets)	17	63%	MAT
(Cedillo-Peláez et al. 2012)	Mexico/ Oaxaca	Dogs (stray)	154	61.7%	ELISA
(Nguyen et al. 2012)	Korea	Dogs (domestic)	553	12.8	LAT
(Hotea et al. 2011a)	Romania/ Caras-Severin County	Dogs	144	68.75%	ELISA
(Hotea et al. 2011a)	Romania/ Arad County	Dogs	152	49.34%	ELISA
(Hotea et al. 2011a)	Romania/ Timis County	Dogs	236	62.29%	ELISA
(Cabezón et al. 2010)	Spain/Majorca	Dogs	46	58.7%	MAT
(Sroka et al. 2010)	Poland/Lublin	Dogs	489	53.6%	DAT
(Santos et al. 2009)	Brazil/ Mato Grosso	Dogs	61	88.5%	IFAT
(Hosseininejad et al. 2009)	Iran/ Shahrekord	Dogs	245	29.79%	IFAT
(Figueredo et al. 2008)	Brazil/ Pernambuco	Dogs	625	57.6%	IFAT
(Lee et al. 2008)	South Korea	Dogs (German Shepherd)	138	46.3%	PCR
(Dubey et al. 2007c)	Vietnam	Dogs	42	50%	MAT
(Dubey et al. 2007b)	USA	Dogs	86	67.4%	MAT
(Jittapalapong et al. 2007)	Thailand/ Bangkok	Dogs	427	9.4%	MLAT
(Javadi et al. 2008)	Iran/Uramia	Dogs	52	84.6%	ELISA

Table 14. Cont.

(Author Name/Year)	Location	Species	Examined Samples (No.)	Positive Samples (%)	Methods
(Brandão et al. 2006)	Brazil/ Minas Gerais	Dogs	25	40.9%	IFAT
(Wanha et al. 2005)	Austria	Dogs	242	26%	IFAT
(Aslantaş et al. 2005)	Turkey/ Ankara	Dogs	116	62.06%	SFDT
(da Silva et al. 2005)	Brazil	Dogs (with nervous signs)	34	26.47%	PCR
(Azevedo et al. 2005)	Brazil/ Paraiba	Dogs	286	45.1%	IFAT
(Mineo et al. 2004)	Brazil/ Minas Gerais	Dogs	369	30.3%	ELISA
(Schatzberg et al. 2003)	New York	Dogs (with encephalitis)	6	50%	PCR
(Cañón-Franco et al. 2003)	Brazil/ Rondonia	Dogs	157	76.4%	IFAT
(Meireles et al. 2004)	Brazil/Sao Paulo	Dogs	200	50.5%	ELISA
(Ali et al. 2003)	Trinidad and Tobago	Dogs	250	32%	MLAT
(Dubey et al. 2003b)	USA	Puppy	20	10%	Immuno-histochemistry
(DeSouza et al. 2003)	Brazil	Dogs (rural)	1244	34.3%	MAT
		Dogs (urban)		19.7%	
(de Brito et al. 2002)	Brazil/Sao Paulo	Dogs (with nervous signs)	80	Ig G–32.5%	IFAT
				Ig M–31.2%	
(Giraldi 2004)	Brazil/ Londrina	Dogs (with nervous signs)	67	82.5%	IFAT

Source: Table by authors.

In Beijing, China, Yu et al. (2008) examined 534 dogs and 335 cats. They observed a variation in prevalence in correlation with the season, noting that in spring, the prevalence in dogs and cats was higher. Of the animals identified as positive for *Toxoplasma gondii* infection, 31.3% of dogs and 25.1% of cats were diagnosed in the spring. In their study, they also observed a higher prevalence in males (23.7% in dogs and 15.1% in cats) than in females (18% in dogs and 12.3% in cats). In terms of age, they observed a higher prevalence in animals over 4 years (31.8% in dogs and 14% in cats) (Yu et al. 2008).

Yarim et al. (2007) from the Samsun School of Medicine, Turkey made a correlation between blood protein concentration and the presence of *T. gondii* infection in naturally infected dogs. The author observed major changes in the serum protein concentration of animals infected with *T. gondii*, correlating with liver alterations, also proven by liver enzyme testing. Thus, he concluded that the determination of serum proteins, by electrophoresis, may be useful in the diagnosis and prognosis of canine toxoplasmosis, as an additional method in addition to another laboratory analysis (Yarim et al. 2007).

A very interesting study was conducted by Bresciani et al. (2009) at the UNESP University of Sao Paulo, Brazil. They demonstrated congenital transmission to *Toxoplasma gondii* in pregnant bitches seropositive and reinfected with this parasite. Twelve HIV-positive pregnant bitches were used, divided into three groups. Those in group I were reinfected by subcutaneous inoculation with tachyzoites, those in group II were reinfected by oral inoculation with oocysts and those in group III were not reinfected; they were used as a control group. Reinfection was performed on day 35 of gestation. All females in groups I and II had loss of gestation or fetal death. From group III, a female gave birth to three apparently healthy but seropositive chicks and one dead chick (Bresciani et al. 2009).

Samples from both adult and newborn females, abortions or stillbirths were tested by several methods: intraperitoneal inoculation of infected blood in mice, serological analysis by IFAT, histopathological examination, pepsin digestion and inoculation in mice and immunohistochemical analysis (Bresciani et al. 2009).

In the study conducted by Bresciani et al. (2009), the parasite could be identified by all applied methods. *T. gondii* was isolated in 17 organs of reinfected females, in 11 organs of females in the control group and in 20 organs of chicks. The most affected organs were the lymph nodes, liver, lungs and kidneys (Bresciani et al. 2009).

Schares et al. (2005) examined the feces of 24,089 dogs in Germany and identified oocysts measuring 9–14 μm. From the samples in which oocysts were identified, inoculations were performed on laboratory animals and two isolates of *T. gondii* were obtained. The two isolates of *T. gondii* could be successfully passed on cell cultures, thus making it possible to characterize them in detail. These data suggest the role of dogs in the dissemination of toxoplasmosis, as possible vectors of the parasite (Schares et al. 2005).

4.6.3. *T. gondii* Infection in Wild Animals

Felids

Congenital toxoplasmosis has been reported in kittens (*Felis rufus*). *Toxoplasma* meningoencephalitis was observed in six-month-old kittens, and *Toxoplasma gondii* was even isolated from tissues in adults (Smith et al. 1995). Lynx is thought to be important in maintaining toxoplasmosis in herbivores in the United States (Dubey et al. 2004a).

Oocysts eliminated by cougars (*Felis concolor*) are thought to be the sources of toxoplasmosis outbreaks in humans through water in Victoria, British Columbia and Canada. Oocysts have been isolated from the feces of cougars collected around water sources (Aramini et al. 1998). Experimental infections resulting in the elimination of oocysts in the feces have been performed in jaguars (*Felis jaguarundi*), ocelots (*Felis pardalis*), lynx (*Lynx rufus*) and cheetah (*Acinonyx jubatus*). In general, these cats are not as important for eliminating oocysts as the domestic cat (Weiss and Kim 2004).

Canids

Acute toxoplasmosis could be observed in the polar fox (*Vulpes lagopus*), the desert fox (*Fennecus zerda*), the gray fox (*Urocyon cinereoargenteus*) and the red fox (*Vulpes vulpes*) (Kelly and Sleeman 2003; Kottwitz et al. 2004; Sørensen et al. 2005). Coinfection of clinical toxoplasmosis with canine scabies virus has been found in gray and red foxes (Kelly and Sleeman 2003). In contrast, signs of clinical toxoplasmosis could not be observed in wolves, coyotes, hyenas or dingo dogs, although the parasite could be identified in gray and red fox, as well as from coyotes (Dubey et al. 2004a; Weiss and Kim 2004; Dubey et al. 2021). In the world, the prevalence of *T. gondii* infection in foxes is varied (Table 15).

Table 15. Seroprevalence of *T. gondii* infection in foxes.

Author Name (Year)	Location	Species	Examined Samples (No.)	Positive Samples (%)	Methods
(Bouchard et al. 2019)	Canada/ Nunavut	Arctic fox (*Vulpes lagopus*)	82	39%	MAT
(Bártová et al. 2018)	Czech Republic	Arctic wolf (*Canis lupus arctos*)	8	62.5%	LAT
(Gerhold et al. 2017)	USA/ Georgia, Tennessee	Gray fox (*Urocyon cinereoargenteus*)	4	50%	MAT
(Waap et al. 2016)	Portugal	Fox (*Vulpes vulpes*)	25	40%	DAT
(Lou et al. 2015)	China	Arctic fox (*Vulpes lagopus*)	1346	8.4%	MAT
(da Silva et al. 2014)	Brazil/Sao Paulo	Hoary fox (*Lycalopex vetulus*)	6	16.6%	MAT
(Verin et al. 2013)	Italy/Pisa	Red fox (*Vulpes vulpes*)	191	53.4%	IFA

Table 15. *Cont.*

Author Name (Year)	Location	Species	Examined Samples (No.)	Positive Samples (%)	Methods
(Jakubek et al. 2012)	Sweden	Red fox (*Vulpes vulpes*)	56	71%	MAT
(De Craeye et al. 2011)	Belgium	Foxes (*Vulpes vulpes*)	304	18.8%	ELISA
(Dubey et al. 2011)	USA/Alaska	Arctic fox (*Vulpes lagopus*)	27	59.3%	MAT
(Aubert et al. 2010)	France	Foxes (*Vulpes vulpes*)	14	64.28%	MAT
(Catenacci et al. 2010)	Brazil	Foxes (*Vulpes vulpes*)	52	19.2%	IFAT
(Dubey 2010)	UAE	Foxes (*Vulpes vulpes*)	32	62.5%	MAT
(Yu et al. 2009)	China	Arctic fox (*Vulpes lagopus*)	103	0.97%	ELISA
(Prestrud et al. 2008)	Svalbard Archipelago	Arctic fox (*Vulpes lagopus*)	11	9.09%	MAT
			178	91.57%	PCR
(Dubey and Pas 2008)	UAE	Blanford's fox (*Vulpes cana*)	12	83.33%	LAT
(Prestrud et al. 2007)	Svalbard Archipelago	Arctic fox (*Vulpes lagopus*)	594	43%	DAT
(Murphy et al. 2007)	Ireland	Foxes (*Vulpes vulpes*)	206	56%	IFAT
(Sobrino et al. 2007)	Spain	Foxes (*Vulpes vulpes*)	102	64.7%	MAT
(Jakubek et al. 2007)	Hungary	Foxes (*Vulpes vulpes*)	337	68%	DAT
(Hůrková and Modrý 2006)	Czech Republic	Foxes (*Vulpes vulpes*)	152	1.32%	PCR
(Wanha et al. 2005)	Austria	Foxes (*Vulpes vulpes*)	94	35%	IFAT

Source: Table by authors.

Bears

Clinical toxoplasmosis could not be observed in bears. Viable forms of *T. gondii* have been isolated from the black bear (*Ursus americanus*), and serological tests have shown the presence of parasitism in the polar bear (*Ursus maritimus*) and the grizzly bear (*Ursus arctos*). Thus, bear meat can be considered a source of infection for carnivores (Rah et al. 2005; Weiss and Kim 2004; Dubey et al. 2010b).

Raccoons

Tests performed on mice (*Procyon lotor*) indicate an increased prevalence in these species (Hancock et al. 2005). Even untreated parasitic forms in rat tissues were isolated, but the clinical form of the disease was not observed. They appear to be resistant to experimental infection (Dubey et al. 2004a; Weiss and Kim 2004).

Squirrels

Acute toxoplasmosis has been identified in the gray squirrel (*Sciurus carolinensis*), the 13-line squirrel (*Citellus tridecemlineatus*) and the Korean squirrel (*Tanias sibericus*) (Carrasco et al. 2006; Dubey et al. 2006c). Parasitic forms could be isolated from the gray squirrel and the giant flying squirrel (*Petaurista petaurista grandis*) (Weiss and Kim 2004).

Rabbits

A fatal form of toxoplasmosis has been found in three domestic rabbits (*Oryctolagus cuniculus*) in the United States. Rabbits died of an acute form of the disease. Similar forms of the disease have been observed in domestic rabbits aged 2 to 18 months from 15 farms in Germany. The necropsy of the 49 rabbits revealed the presence of a generalized toxoplasmosis with granulomatous and necrotizing lesions in the spleen, liver, lungs and lymph nodes. Experimental infections have also been reported in rabbits. They were inoculated orally and subcutaneously with *T. gondii* oocysts (10,000 oocysts/rabbit) and all developed fatal forms of toxoplasmosis, or the lesions were so severe that euthanasia was required (Weiss and Kim 2004). Worldwide, the prevalence of *T. gondii* infection in rabbits is varied (Table 16).

Table 16. Seroprevalence of *T. gondii* infection in rabbits.

(Author Name/Year)	Location	Species	Examined Samples (No.)	Positive Samples (%)	Methods
(Waap et al. 2016)	Portugal	European rabbit (*Oryctolagus cuniculus*)	36	2.8%	DAT
(Mecca et al. 2011)	Brazil/Sao Paulo	Wild rabbits (*Lepus europaeus*) (meat)	74	1.35%	ELISA
(Ashmawy et al. 2011)	Egypt	Wild rabbits (*Lepus europaeus*)	194	11.34%	IHAT

Table 16. *Cont.*

(Author Name/Year)	Location	Species	Examined Samples (No.)	Positive Samples (%)	Methods
(Aubert et al. 2010)	France	Wild rabbits (*Lepus europaeus*)	3	0%	MAT
(Bártová et al. 2010)	Czech Republic	Wild rabbits (*Lepus europaeus*)	333	21%	IFAT
	Slovakia		209	6%	
	Austria		383	13%	
(Jeklova et al. 2010)	Czech Republic	Domestic rabbits (*Oryctolagus cuniculus*)	500	19.2%	ELISA
(García-Bocanegra et al. 2010d)	Spain	Domestic rabbits (*Oryctolagus cuniculus*)	85	11.9%	MAT
(Harfoush and Tahoon 2010)	Egypt	Domestic rabbits (*Oryctolagus cuniculus*)	118	37.5%	IHAT
(Künzel et al. 2008)	Austria/ Vienna	Domestic rabbits (*Oryctolagus cuniculus*)	157	5.09%	IFAT
(Hughes et al. 2008)	UK/ Malham	Domestic rabbits (*Oryctolagus cuniculus*)	57	68.4%	PCR
(Figueroa-Castillo et al. 2006)	Mexico	Domestic rabbits (*Oryctolagus cuniculus*)	286	26.9%	ELISA
(Almería et al. 2004)	Spain	Domestic rabbits (*Oryctolagus cuniculus*)	456	14.2%	MAT

Source: Table by authors.

In another study, hares (*Lepus europaeus*) also developed the fatal form of the disease after inoculation with only 10 oocysts and all died from day 8 post-inoculation to day 19 (Sedlák et al. 2000). Experimental infection of mountain rabbits (*Lepus timidus*) with 50 *T. gondii* oocysts resulted in multiple lesions of the mesenteric lymph nodes and liver. Histologically, there have been extensive areas of necrosis in the small intestine, mesenteric lymph nodes, liver and other internal organs (Gustafsson et al. 1997a, 1997b).

Skunks

Genotype III of *Toxoplasma gondii* could be isolated from three of six asymptomatic striped scorpions (*Mephitis mephitis*) in Mississippi (Dubey et al. 2004e). Two of the three isolated forms were pathogenic to rodents, even those characterized molecularly as part of genotype III, which is considered avirulent for rodents (Weiss and Kim 2004).

Martens

Toxoplasma gondii was isolated by PCR from the brains and skeletal muscles of young martens (*Martes pennanti*) on a Maryland farm. Clinically, these animals showed encephalitis which, initially, could not be associated with *T. gondii* infection (Gerhold et al. 2005).

Beavers

The parasite was also isolated from beavers (*Castor canadensis*), but no clinical signs of the disease were observed. By direct agglutination, of the 62 samples examined, 6 (i.e., 10%) were found to be positive (Jordan et al. 2005).

Insectivores

There are few data on toxoplasmosis in insectivores. In the Czech Republic, out of 578 insectivores examined, less than 1% of them tested positive for the Sabin–Feldman staining test. A positive case is known in a male mole (*Talpa europaea*) from Germany (Weiss and Kim 2004).

Bats

No clinical cases of toxoplasmosis have been reported in bats, but the parasite could be isolated from Vesperzilio pipistrellus bats and red bats (*Nyctalus noctula*) from Alma-Ata, Kazakhstan, USSR (Weiss and Kim 2004).

Cervidae

Toxoplasma gondii has been identified in North American deer. Consumption of venison has been associated with the clinical form of the disease in humans (Ross et al. 2001).

Naturally infected deer did not show clinical signs, but the parasite could be isolated from tissues harvested from white-tailed deer (*Odocoileus virginianus*) and large-eared deer (*Odocoileus hemionus*) (Dubey et al. 2004a). Fetuses from HIV-positive mothers of *Odocoileus virginianus* and *Odocoileus hemionus* did not show anti-*Toxoplasma* antibodies, which proves that in these two breeds of deer, antibodies are not transmitted from mother to fetus. Acute or fatal forms of the disease can occur in experimental infections.

Congenital toxoplasmosis has been identified in other deer, such as reindeer (*Rangifer tarandus*). They can present enteritis and even death, as demonstrated after an experimental oral inoculation with oocysts (Dubey et al. 2002a; Lindsay et al. 2005). *Toxoplasma gondii* has also been isolated from roe deer (European roe) (*Capreolus capreolus*) (Weiss and Kim 2004). Globally, the prevalence of *T. gondii* infection in these species is varied (Table 17).

Table 17. Seroprevalence of *T. gondii* infection in Cervidae.

(Author Name/Year)	Location	Species	Examined Samples (No.)	Positive Samples (%)	Methods
(Remes et al. 2018)	Estonia	Moose (*Alces alces*)	463	23.97%	DAT
(Waap et al. 2016)	Portugal	Red deer (*Cervus elaphus*)	14	21.4%	DAT
(Malmsten et al. 2011)	Sweden	Roe-deer (*Capreolus capreolus*)	199	34%	ELISA
		Moose (*Alces alces*)	417	20%	
(De Craeye et al. 2011)	Belgium	Roe-deer (*Capreolus capreolus*)	73	52%	ELISA
		Red-deer (*Cervus elaphus*)	7	0%	
(Panadero et al. 2010)	Spain/ Galicia	Roe-deer (*Capreolus capreolus*)	160	13.7%	DAT
(Jokelainen et al. 2010)	Finland/ Helsinki	Moose (*Alces alces*)	1215	9.6%	DAT
		Key-deer (*Odocoileus virginianus*)	135	26.7%	
		Roe-deer (*Capreolus capreolus*)	17	17.6%	
(Aubert et al. 2010)	France	Roe-deer (*Capreolus capreolus*)	33	36.3%	MAT
(Dubey et al. 2009)	USA/Iowa	Key-deer (*Odocoileus virginianus*)	170	53.5%	MAT
	USA/ Minnesota		62	32.2%	

Table 17. Cont.

(Author Name/Year)	Location	Species	Examinated Samples (No.)	Positive Samples (%)	Methods
(Dubey et al. 2008b)	Iowa	Key-deer (*Odocoileus virginianus*)	84	64.2%	MAT
	Minnesota		27	55.55%	
(Dubey et al. 2008c)	USA/ Washington	Black-tailed Deer (*Odocoileus hemionus columbianus*)	43	32.55%	DAT
		Mule deer (*Odocoileus hemionus*)	42	0%	
(Gamarra et al. 2008)	Spain	Roe-deer (*Capreolus capreolus*)	278	39.2%	MAT
(Gauss et al. 2006)	Spain	Red-deer (*Cervus elaphus*)	441	15.6%	MAT
		European fallow deer (*Dama dama*)	79	24%	
		Roe-deer (*Capreolus capreolus*)	33	21.8%	
		Pyrenean chamois (*Rupicapra pyrenaica*)	10	20%	
(Omata et al. 2005b)	Japan/ Hokkaido	Sika deer (*Cervus nippon*) (brain)	120	0%	PCR
(Lindsay et al. 2005)	Nebraska	Mule deer (*Odocoileus hemionus*)	89	35%	MAT
		Key-deer (*Odocoileus virginianus*)	14	43%	

Source: Table by authors.

Other wild ruminants

Elk (*Cervus canadensis*) appears to be resistant to the clinical form of the disease, but the parasite could have been isolated from several tissues, suggesting that elk may be a source of infection for humans. *Toxoplasma gondii* has also been isolated from naturally infected antelopes. Animals experimentally infected with oocysts showed acute and deadly toxoplasmosis (Weiss and Kim 2004).

Toxoplasmic encephalitis was also seen in a 4-month-old Rocky Mountain goat (*Ovis canadensis canadensis*). The parasite could also be isolated from naturally infected moose (*Alces alces*) (Baszler et al. 2000; Weiss and Kim 2004).

Otters and other marine mammals

In the early 1990s, toxoplasmosis was considered a major cause of see otter (*Enhydra lutris nereis*) mortality. The predominant symptom was encephalitis (Kreuder et al. 2003). This has been difficult to explain because otters are not considered to consume species that are common intermediate hosts, and their marine habitat keeps them away from cats. The decisive evidence that *T. gondii* is responsible for otter death was the identification of viable parasitic forms in otter tissues. Additionally, the parasite isolated from sea otters appears to have been able to grow, and cats that have consumed infested tissues have eliminated oocysts. Genotype II was initially isolated, but a few years later a new genotype, called genotype X, was isolated (Miller et al. 2004, 2008a).

It has been thought that oocysts eliminated by cats living on the Pacific coast end up being eaten by otters through paratenic hosts. *Toxoplasma gondii* oocysts sporulate in seawater and can remain infectious for at least 6 months (Putignani et al. 2011). This has been demonstrated by isolating viable forms of the parasite, or at least DNA, from the body of experimentally infected shells and maintained in natural habitat (Arkush et al. 2003). *Toxoplasma gondii* kills other marine mammals off the Pacific coast. Genotype × was also isolated from Harbor seals (*Phoca vitulina*) or California sea lions (*Zalophus californianus*) (Conrad et al. 2005).

Toxoplasmosis is also common in dolphins. The congenital form has been described in Indo-Pacific bottlenose dolphins (*Tursiops aduncus*). Disseminated toxoplasmosis with transplacental transmission has also been identified in a pregnant female of Risso's dolphin (*Grampus griseus*). The acute form of the disease has been described in various dolphins: *Sousa chinensis, Stenella longirostris* and *Tursiops truncatus* (Weiss and Kim 2004). Toxoplasmosis has also been identified in white dolphins (*Delphinapterus leucas*), Californian sea lions (*Zalophus californianus*), fur seals (*Callorhinus ursinus*), sea elephants (*Mirounga angustirostris*), Hawaiian seals (*Monachus schaulnslandt*) or Indian manatee (*Trichechus manatus*) (Esmerini et al. 2010; Massie et al. 2010).

In the case of experimental infections in the gray seal (*Halichoerus grypus*) with over 10,000 oocysts, no acute toxoplasmosis occurred, but the parasite could be isolated from the brain and muscles (Gajadhar et al. 2004). Globally, the prevalence of *T. gondii* infection in these species is varied (Table 18).

Table 18. Seroprevalence of T. gondii infection in marine mammals.

(Author Name/Year)	Location	Species	Examinated Samples (No.)	Positive Samples (%)	Methods
(van de Velde et al. 2016)	Scottish coastline – Atlantic Ocean and North Sea	Long-finned pilot whale (*Globicephala melas*)	6	83%	MAT
				60%	ELISA
				60%	IFAT
		Risso's dolphin (*Grampus griseus*)	4	50%	MAT
				25%	ELISA
				25%	IFAT
		Short-beaked common dolphin (*Delphinus delphis*)	9	56%	MAT
				11%	ELISA
				11%	IFAT
		White-beaked dolphin (*Lagenorhynchus albirostris*)	4	75%	MAT
				50%	ELISA
				25%	IFAT
		Gray seal (*Halichoerus grypus*)	12	25%	MAT
				11%	ELISA
				0%	IFAT
		Harbor seal (*Phoca vitulina*)	14	29%	MAT
				18%	ELISA
				18%	IFAT
		Long-finned pilot whale (*Globicephala melas*)	5	60%	MAT
				20%	ELISA
				20%	IFAT
		European otter (*Lutra lutra*)	34	47%	MAT
				44%	ELISA
				53%	IFAT
(Rengifo-Herrera et al. 2012)	South Shetland Islands and Antarctic Peninsula	Antarctic pinnipeds	211	13.3%	DAT
(Cabezón et al. 2011)	N-E Atlantic coast	Gray seal (*Halichoerus grypus*)	47	23.4%	MAT
		Harbor seal (*Phoca vitulina*)	56	5.4%	

Table 18. *Cont.*

(Author Name/Year)	Location	Species	Examined Samples (No.)	Positive Samples (%)	Methods
(Deem et al. 2010)	Ecuador/ Galapagos	Galapagos penguin (*Spheniscus mendiculus*)	298	2.3%	MAT
		Flightless cormoran (*Phalacrocorax harrisi*)	258	1.9%	
(Forman et al. 2009)	UK	Marine mammals (various)	101	7.9%	SFDT
(Miller et al. 2008b)	USA	Sea otter (*Enhydra lutris nereis*) (brain)	1	100%	Histology
		Sea otter (*Enhydra lutris nereis*) (serum)	35	16%	IFAT
(Sundar et al. 2008)	USA/ California, Washington	Sea otter (*Enhydra lutris*)	37	39 isolates	PCR
(Prestrud et al. 2007)	Svalbard Archipelago	Walrus (*Odobenus rosmarus*)	17	6%	DAT
(Conrad et al. 2005)	USA	Sea otter (*Enhydra lutris nereis*) (dead bodies)	305	52%	Serology
		Sea otter (*Enhydra lutris nereis*)	257	38%	
(Measures et al. 2004)	Canada	Harp seal (*Phoca groenlandica*)	112	0%	MAT
		Hooded seal (*Cystophora cristata*)	60	2%	
		Gray seal (*Halichoerus grypus*)	122	9%	
		Harbor seal (*Phoca vitulina*)	34	9%	
(Murata et al. 2004)	Japan	Wild and captive cetaceans	59	11.9%	LAT and IFAT
(Cabezón et al. 2004)	Spain	Striped dolphin (*Stenella coeruleoalba*)	36	11.11%	MAT
		Atlantic bottlenose dolphin (*Tursiops truncatus*)	7	57.14%	

Table 18. *Cont.*

(Author Name/Year)	Location	Species	Examined Samples (No.)	Positive Samples (%)	Methods
(Dubey et al. 2003c)	USA	Marine mammals (dead bodies)	115	77%	MAT
		Sea otter (*Enhydra lutris*)	30	60%	
		Harbor seal (*Phoca vitulina*)	311	16%	

Source: Table by authors.

New World monkeys

Toxoplasmosis can be considered a problem for New World monkeys. Several data on acute toxoplasmosis have referred to the Common squirrel monkey (*Saimiri sciureus*) and Golden lion tamarin (*Leontopithecus rosalia*) (Epiphanio et al. 2003). The Panamanian squirrel monkey and Gray-bellied night monkey (*Aotus lemurinus*) are also very susceptible to experimental infection with tissue cysts and develop an acute, fatal form of the disease (Furuta et al. 2001). Other species of monkey diagnosed with toxoplasmosis are shown in Table 19.

Table 19. Monkeys species that developed clinical toxoplasmosis (by Weiss and Kim 2004).

No.	Monkey Species
1	Saguinus oedipus
2	Saguinus midas midas
3	Saguinus midas niger
4	Saguinus imperator
5	Saguinus labiatus
6	Leontopithecus chrysopygus
7	Leontopithecus chrysomelas
8	Leontopithecus rosalia
9	Saimiri sciureus
10	Callithrix pygmaea
11	Callithrix jacchus
12	Callithrix pencillata
13	Pithecia pithecia
14	Aotus trivirgatus
15	Alouatta fusca
16	Lagothrix lagotricha

Source: Reprinted from (Weiss and Kim 2004), used with permission.

Old World monkeys

In Old World monkeys, toxoplasmosis is quite rare (Table 20). An associated case of nervous system toxoplasmosis with immunodeficiency virus (AIDS), resulting in encephalomyelitis, has been cited in a macaque (*Macaca sylvana*). These types of monkeys (*Macaca mulatta, Macaca arctoides*) have been/are used in experimental infections to study human congenital toxoplasmosis and recurrent retinochoroiditis (*Macaca fascicularis*) (Weiss and Kim 2004).

Table 20. Seroprevalence of *T. gondii* infection in varios species of wild mammals and birds.

(Author Name/Year)	Location	Species	Examinated Samples (No.)	Positive Samples (%)	Methods
(Bártová et al. 2021)	Czech Republic	Ostriches (*Struthio camelus*)	409	36%	LAT
(Ferreira et al. 2019)	Africa	African lions	15	100%	ELISA
		Spotted hyenas	41	93%	
(Bártová et al. 2018)	Czech Republic (Zoo)	Accipitriformes	17	71%	LAT
		Anseriformes	3	66.67%	
		Pelecaniformes	18	6%	
		Psittaciformes	9	22%	
		Sphenisciformes	4	50%	
		Canidae	46	59%	
		Felidae	71	70%	
(Bártová et al. 2018)	Czech Republic (Zoo)	Ursidae	6	83.3%	LAT
		Bovidae	446	27.8%	
		Camelidae	34	32.4%	
		Cervidae	70	45.7%	
		Giraffidae	13	23.1%	
		Hippopotamidae	3	0	
(Bártová et al. 2018)	Czech Republic (Zoo)	Suidae	5	20%	LAT
		Macropodidae	4	50%	
		Equidae	123	27.6%	
		Rhinocerotidae	18	16.7%	
		Primates	47	27.7%	
		Cercopithecidae	15	13.3%	
		Hominidae	10	30%	
		Elephantidae	11	36.4%	
		Rodentia	24	12.5%	

Table 20. Cont.

(Author Name/Year)	Location	Species	Examined Samples (No.)	Positive Samples (%)	Methods
(Dubey et al. 2017)	USA/ Pennsylvania	North American elk (*Cervus canadensis*)	317	69.7%	MAT
(da Silva and Langoni 2016)	Brazilia	Ostriches (*Struthio camelus*)	344	11.05%	MAT
(Waap et al. 2016)	Portugal	Cat (*Felis catus*)	79	39.2%	DAT
		Wildcat (*Felis silvestris*)	6	83.3%	
		European badger (*Meles meles*)	6	0	
		European otter (*Lutra lutra*)	2	0	
(Waap et al. 2016)	Portugal	Stone marten (*Martes foina*)	6	66.7%	DAT
		European polecat (*Mustela putorius*)	3	33.3%	
		Egyptian mongoose (*Herpestes ichneumon*)	34	85.3%	
		Common genet (*Genetta genetta*)	17	47.1%	
(Chen et al. 2015)	Taiwan	Columbiformes	102	9.8%	MAT
		Passeriformes	45	37.78%	
		Galliformes	2	0	
		Cuculiformes	2	0	
		Piciformes	2	0	
		Gruiformes	1	0	
		Charadriiformes	2	0	
		Anseriformes	1	100%	
		Pelecaniformes	30	36.67%	
		Strigiformes	111	22.52%	
		Accipitriformes	93	29.34%	
		Falconiformes	2	50%	

Table 20. *Cont.*

(Author Name/Year)	Location	Species	Examinated Samples (No.)	Positive Samples (%)	Methods
(Cedillo-Peláez et al. 2011)	Mexico	Monkeys	2	100%	PCR
(Fornazari et al. 2011)	Brazil/Sao Paulo	White-eared opossum (*Didelphis albiventris*)	72	5.5%	MAT
(Dărăbuş et al. 2011b)	Romania/ Timisoara Zoo	Red deer (*Cervus elaphus*)	7	28.57%	ELISA
		Roe deer (*Capreolus capreolus*)	4	75%	
(Dărăbuş et al. 2011b)	Romania/ Timisoara Zoo	Reindeer (*Rangifer tarandus*)	4	50%	ELISA
		Guanaco (*Lama guanicoe*)	2	100%	
		Goat (*Capra aegagrus hircus*)	9	44.44%	
		Shetland Pony (*Equus caballus*)	5	40%	
		European rabbit (*Oryctolagus cuniculus*)	12	41.66%	
		Red-necked wallaby (*Macropus rufogriseus*)	2	0	
		Patagonian mara (*Dolichotis patagonum*)	3	33.33%	
		Brown bear (*Ursus arctos*)	3	100%	
		Raccoon (*Procyon lotor*)	4	100%	
		Japanese macaque (*Macaca fuscata*)	5	40%	
		Patas monkey (*Erythrocebus patas*)	3	33.33%	

Table 20. *Cont.*

(Author Name/Year)	Location	Species	Examinated Samples (No.)	Positive Samples (%)	Methods
		Lion (*Panthera leo*)	3	100%	
		Wildcat (*Felis sylvestris*)	2	100%	
		Domestic cat (from Zoo)	1	100%	
(Aubert et al. 2010)	France	European mouflon (*Ovis gmelini musimon*)	7	14.28%	MAT
		Mallard (*Anas platyrhynchos*)	2	50%	
(Gondim et al. 2010)	Brazil	House sparrow (*Passer domesticus*)	293	1.02%	HAT
(Dubey 2010)	United Arab Emirates	Wild Felids	84	83.33%	MAT
(Parameswaran et al. 2010)	Australia	Marsupials	46	26.08%	PCR
(García-Bocanegra et al. 2010d)	Spain	Iberian lynx (*Lynx pardinus*)	129	62.8%	MAT
(Ullmann et al. 2010)	Southern Brazil	Wild Felids	57	66.67%	MAT
(Moré et al. 2010)	Argentine	Kangaroos (*Macropus rufus* and *Macropus giganteus*)	2	100%	PCR
(Yai et al. 2008)	Brazil	Capybara (*Hydrochoerus hydrochaeris*)	36	44.5%	PCR
(Carme et al. 2009)	France	Common squirrel monkey (*Saimiri sciureus*)	35	57.14%	PCR
(Parameswaran et al. 2009)	Australia	Kangaroos	219	15.5%	ELISA
(Basso et al. 2009)	Argentine	Meerkat (*Suricata suricata*) from Zoo	3	33.3%	Inoculation on mice

(Author Name/Year)	Location	Species	Examined Samples (No.)	Positive Samples (%)	Methods
(Salant et al. 2009b)	Israel	Common squirrel monkey (*Saimiri sciureus*)	24	79.16%	MAT
(Bermúdez et al. 2009)	Spain	Red-necked wallaby (*Macropus rufogriseus*)	1	----	Identification of tissue cysts in the my-ocardium
(da Silva et al. 2008)	Brazil/Sao Paulo	Nine-banded armadillo (*Dasypus novemcinctus*)	31	12.9%	MAT
(Liu et al. 2008)	China/ Qinghai	Yac (*Bos grunniens*)	946	11.8%	ELISA
(Hill et al. 2008)	Australia/ Sydney	Common brushtail possum (*Trichosurus vulpecula*) (from Zoo)	126	4.8%	MAT
		Dingo (*Canis familiaris dingo*)	23	67%	
		Chimpanzee (*Pan troglodites*)	23	22%	
(Sobrino et al. 2007)	Spain	Iberian lynx (*Lynx pardinus*)	27	81.5%	MAT
		Wildcat (*Felis silvestris*)	6	50%	
		Wolfs (*Canis lupus*)	32	46.9%	
		European badger (*Meles meles*)	37	70.3%	
		Beech marten (*Martes foina*)	20	85%	

(Author Name/Year)	Location	Species	Examined Samples (No.)	Positive Samples (%)	Methods
(Sobrino et al. 2007)	Spain	European pine marten (*Martes martes*)	4	100%	MAT
		Eurasian otter (*Lutra lutra*)	6	100%	
		European polecat (*Mustela putorius*)	4	100%	
		Ferret (*Mustela putorius furo*)	1	100%	
		Common genet (*Genetta genetta*)	21	61.9%	
		Egyptian mongoose (*Herpestes ichneumon*)	22	59.1%	
(Yabsley et al. 2007)	USA/ Catherine Island	Ring-tailed lemur (*Lemur catta*)	52	5.8%	MAT
(Basso et al. 2007)	Argentine	Red-necked wallaby (*Macropus rufogriseus*)	4	75%	MAT
(Dangolla et al. 2006)	Turkey /Sri Lanka	Elephants (bred in captivity)	45	32%	MAT
(Hůrková and Modrý 2006)	Czech Republic	Martens	61	4.92%	PCR
(Anwar et al. 2006)	UK	European badger (*Meles meles*)	90	70%	LAT
(Ryser-Degiorgis et al. 2006)	Sweden	Eurasian lynx (*Lynx linx*)	207	75.4%	DAT
(Mucker et al. 2006)	USA/ Pennsylvania	Bobcat (*Lynx rufus rufus*)	131	83%	MAT
(Eymann et al. 2006)	Australia/ Sydney	Common brushtail possum (*Trichosurus vulpecula*)	142	6.3%	MAT

Table 20. Cont.

(Author Name/Year)	Location	Species	Examined Samples (No.)	Positive Samples (%)	Methods
(Thiangtum et al. 2006)	Thailand	Felids (bred in captivity)	136	15.4%	LAT
(Rajkhowa et al. 2006)	India	Gayal (*Bos frontalis*)	104	28%	MAT
(Sadrebazzaz et al. 2006)	Iran/ Mashhad	Dromedary (*Camelus dromedarius*)	120	4.16%	IFAT
(Garcia et al. 2005)	Brazil/ Parana	New World Monkeys: Capuchin (*Cebus* spp.)	43	30.2%	MAT
		Black howler (*Alouatta caraya*)	17	17.6%	
(Hove et al. 2005a)	Zimbabwe	Lion (*Panthera leo*)	26	92%	MAT
		Nyala (*Tragelaphus angasii*)	10	90%	
		Greater kudu (*Tragelaphus strepsiceros*)	10	20%	
		Giraffe (*Giraffa camelopardalis*)	10	10%	
		African bush elephant (*Loxodonta africana*)	20	10%	
		Common ostrich (*Struthio camelus*)	50	48%	
(Hancock et al. 2005)	USA/Virginia	Raccoon (*Procyon lotor*)	256	84.4%	MAT
(Hartley and English 2005)	Australia	Common wombat (*Vombatus ursinus*)	23	26.1%	LAT
(Marobin et al. 2004)	Brazil/Rio Grande do Sul	Greater rhea (*Rhea americana*)	74	8.1%	IHA

Table 20. Cont.

(Author Name/Year)	Location	Species	Examined Samples (No.)	Positive Samples (%)	Methods
(Kajerová et al. 2003)	Czech Republic	Budgerigar (*Melopsittacus undulatus*)	53	100%	Experimental infection and LAT
(Piasecki et al. 2004)	Poland	Hawks	28	100%	LAT
(Zhang et al. 2004)	China/ Yuanjiang	Reed vole (*Microtus fortis*)	124	29%	MAT
(Vitaliano et al. 2004)	Brazil	Maned wolf (*Chrysocyon brachyurus*) (bred in captivity)	59	74.6%	ELISA
(DeFeo et al. 2002)	Rhode Island	Wild rodents	756	0.8%	MAT
(Silva et al. 2001)	Brazil	Felids (total)	865	54.6%	MAT
		Jaguarundi (*Herpailurus yagouaroundi*)	99	45.9%	
		Ocelot (*Leopardus pardalis*)	168	57.7%	
		Oncilla (*Leopardus tigrinus*)	131	51.9%	
		Margay (*Leopardus wiedii*)	63	55.5%	
		Pampas Cat (*Oncifelis colocolo*)	8	12.5%	
(Silva et al. 2001)	Brazil	Geoffroy's cat (*Oncifelis geoffroyi*)	12	75%	MAT
		Jaguar (*Panthera onca*)	212	63.2%	
		Cougar (*Puma concolor*)	172	48.2%	

Table 20. *Cont.*

(Author Name/Year)	Location	Species	Examinated Samples (No.)	Positive Samples (%)	Methods
(Tuntasuvan et al. 2001)	Thailand	Indian elephant (*Elephus maximus indicus*)	156	45.5% 25.6%	MAT LAT
(Kutz et al. 2000)	Canada	Musk ox (*Ovibos moschatus*)	203	6.4%	MAT
(Dubey et al. 2000)	Canada	Ostrich (*Struthio camelus*)	973	2.9%	MAT
(Hove and Dubey 1999)	Zimbabwe	Eland (*Taurotragus oryx*)	19	36.8%	MAT
		Sable antelope (*Hippotragus niger*)	67	11.9%	
		White rhinoceros (*Ceratotherium simus*)	2	50%	
		African buffalo (*Syncerus caffer*)	18	5.6%	
		Blue wildebeest (*Connochaetas taurinus*)	69	14.5%	
		African bush elephant (*Loxodonta africana*)	19	10.5%	
(Mitchell et al. 1999)	USA/ Illinois	Raccoon (*Procyon lotor*)	459	49%	Serology
(Aramini et al. 1998)	Canada/ Vancouver	Cougar (*Felis concolor vancouverensis*)	12	92%	MAT

Table 20. *Cont.*

(Author Name/Year)	Location	Species	Examined Samples (No.)	Positive Samples (%)	Methods
(Hill et al. 1998)	USA/Iowa	Raccoon (*Procyon lotor*)	885	15%	MAT
		Striped skunk (*Mephitis mephitis*)	81	47%	
		Virginia opossum (*Didelphis virginiana*)	53	23%	
		Stray cats	20	80%	
(Zarnke et al. 1997)	Alaska	Brown bear (*Ursus arctos*)	892	25%	MAT
(Lindsay et al. 1996)	Texas	Coyote (*Canis latrans*)	52	62%	Serology
(Dubey et al. 1995b)	USA/ Illinois	Raccoon (*Procyon lotor*)	188	67%	MAT
		Striped skunk (*Mephitis mephitis*)	18	38.9%	
		Virginia opossum (*Didelphis virginiana*)	128	22.7%	
		Brown rat (*Rattus novegicus*)	95	6.3%	
		Deer mice (*Peromyscus*)	61	4.9%	
		House mouse (*Mus musculus*)	1243	2.1%	
(Webster 1994)	UK	Brown rat (*Rattus novegicus*)	671	35%	Serology
(Johnson et al. 1990)	Australia	Dingo (*Canis familiaris dingo*)	62	10%	DAT

Source: Table by authors.

North American marsupials

Toxoplasma gondii has been isolated from opossums (*Didelphis virginiana*) in North America. They are more resistant to experimental infections than marsupials in Australia (Weiss and Kim 2004). Prevalence of toxoplasmosis in these species are shown in Table 20.

Australian marsupials

Toxoplasma gondii infection is frequently cited as a cause of death in marsupials in Australia and New Zealand. Many of the data presented regarding the *Toxoplasma* infection in marsupials belong to zoos. These animals develop toxoplasmosis even in the absence of cats, which proves their high susceptibility to this disease.

Canfiel et al. (1990, quoted by Weiss and Kim 2004) describe some of the clinical signs, necropsy lesions and histopathological changes encountered in various marsupials: 43 macropods, 2 marsupial bears (wombat—*Vombatus ursinus*), 2 Koala bears (*Phascolarctos cinereus*), 6 opossums, 15 *Dasyuridae*, 2 numbats (*Myrmecobius fasciatus*), 8 *Peramelemorphia* and 1 *Macrotisnlagotis* (Weiss and Kim 2004). The animals either died suddenly, without clinical form, or showed various signs associated with respiratory, neurological or enteric disorders. Congenital toxoplasmosis was found in some kangaroos (*Macropus fuliginosus melanops*), and the parasite could be isolated in the tissues of an 82-day-old chick. In the case of two Koala bears (*Phascolarctos cinereus*), which died on a reservation in Sydney, Australia, the parasite was isolated from the heart, kidneys, liver, lungs, lymph nodes, spleen, stomach and small intestine (Weiss and Kim 2004).

Experimental infections support claims that Australian marsupials are highly susceptible to toxoplasmosis. Some marsupial rodents (*Perameles gunnii*) develop acute toxoplasmosis after ingestion of 100 oocysts and die after 15–17 days post-infection, showing lesions specific to acute toxoplasmosis (Bettiol et al. 2000). Researchers even believe that in these species (*Perameles gunnii*), toxoplasmosis is one of the causes of the reduction in the amount of wildlife. Other species (*Macropus eugenii*), inoculated with 500, 1000 or 10,000 oocysts, die after 9–15 days after infection (Weiss and Kim 2004). Worldwide, the prevalence of *T. gondii* infection in marsupials varies a lot (Table 20).

Wild animals from Africa

Little is known about mammalian toxoplasmosis in Africa. The clinical form of the disease could not be observed in elephants, hippos, giraffes, gas, wildebeest, South African antelopes, chimpanzees, baboons, orangutans or gorillas, nor could the parasite be isolated. *Toxoplasma gondii* could only be isolated from domestic chickens and ducks in Africa (Weiss and Kim 2004). The prevalence of toxoplasmosis in these species is shown in Table 20.

Wild birds

Toxoplasma gondii has been isolated from the heart and breast of birds of prey. The necropsy lesions described were necrotizing myocarditis in bald eagles (*Haliaeetus leucocephalus*) in New Hampshire and severe hepatitis in captive owls (*Strix varia*) in Quebec, Canada (Dubey et al. 2002c). In the case of three red-tailed hawks (*Buteo jamaicensis*) experimentally infested with tissue cysts, no clinical signs were observed, although the parasite could be isolated from all three individuals. Additionally, in the case of owls (*Bubo virginianus, Strix varia, Asio otus*) which were experimentally infested, the clinical disease did not manifest itself, but *T. gondii* was identified in these birds. In contrast, the parasite could not be isolated from an experimentally infested sparrow hawk (*Falco sparverius*) (Szabo et al. 2004).

Viable forms of the parasite could be isolated from the heart of 8 of 16 turkeys (*Meleagris gallopavo*) in Alabama, and fatal forms of the disease have been observed in wild turkeys in Georgia and West Virginia. Genotype III of *Toxoplasma gondii* has been identified in one of four Canadian geese (*Brante canadensis*) in Mississippi (Dubey et al. 2004b, 2004c; Weiss and Kim 2004) (Table 20).

Fish, reptiles and amphibians

Toxoplasmosis does not occur in these species. Reptiles can be made susceptible to *T. gondii*, but this would mean that they would be experimentally infected and maintained at a temperature of 37–40 °C (Weiss and Kim 2004).

4.6.4. *T. gondii* Infection in Zoos

The presence of toxoplasmosis in Zoos is a management problem, both due to the impossibility of controlling the elimination of oocysts by cats in these areas, and due to the accidental access of stray cats.

Among the mammal species in Zoos that frequently develop toxoplasmosis are Australian marsupials, New World monkeys and Arboreal monkeys, lemurs and Pallas cats. The lesions identified were characteristic of acute toxoplasmosis, but the most severely affected organs were the lungs, liver and spleen. Sporadic cases of acute toxoplasmosis have been reported in some antelopes (*Madoqua guentheri smithi*), meerkats (*Suricata suricatta*) and porcupines (*Coendou mexicanus*). A case of toxoplasmic abortion has been reported in the Green Ox (*Ovibos moschatus wardi*) in Greenland, but abortion and neonatal death have also been reported in some captive antelopes (*Boselaphus tragocamelus*). Additionally, in the antelope (*Saiga tatarica*), in an adult female, the fatal form of the disease was also identified. *T. gondii* was isolated, and confirmed by PCR, from the liver, lungs, spleen, kidneys and intestines (Weiss and Kim 2004).

Acute toxoplasmosis has been described in captive gazelles (*Gazella cuvieri, Gazella leptoceros, Gazella dama, Litocranius walleri*) in Zoos in North America. These infections were spread, and the most affected organs were the liver, lungs, lymph nodes, adrenal glands, spleen, intestines and brain. Toxoplasmosis has also been found in zoological birds, such as canaries. Genotype III was isolated in five out of five black-winged parrots (*Eos cyanogenia*) following the diagnosis of acute toxoplasmosis in a South Carolina avian (Weiss and Kim 2004).

Management and administrative programs in Zoos could be useful for the prevention of toxoplasmosis in mammalian and bird species with a high susceptibility to this disease. Cats should not be fed fresh, unfrozen meat due to the possibility of tissue cyst infestation. Meat should be frozen and thawed before administration to cats, as tissue cysts are destroyed at freezing temperature. The access of common cats to Zoos should be strictly controlled, due to the possibility of them spreading oocysts. Species with a high susceptibility to toxoplasmosis should not be housed near cats (Figure 15).

Figure 15. *Toxoplasma* risk factors for animals from Timisoara Zoo, Romania. Source: Photos by authors.

Outdoor birds also pose a risk of spreading toxoplasmosis by ingesting oocysts, either directly from cat feces or by consuming vector hosts (insects, cockroaches—kitchen) and eliminating them inside Zoos (Weiss and Kim 2004) (see Table 20).

4.6.5. *T. gondii* Infection in Endangered Species

Toxoplasmosis can also adversely affect endangered species of mammals and birds. The Alala crow (*Hawaiian crow, Corvus hawaiiensis*) is an endangered species of which only 25 specimens remained in captivity and in the wild in 2000. Unfortunately, this is one of the species very susceptible to fatal toxoplasmosis, and it develops the disease after being reintroduced into the wild. Toxoplasmosis seems to be a major problem facing the program that aims to introduce this species back into the wild. Groups of monkeys—*Leontopithecus rosalia*—bred in captivity in North America and Europe appear to frequently develop the acute and fatal form of the disease. These tree monkeys are on the verge of extinction and are being raised in captivity for their eventual release back into the wild, with the hope of saving the species. Because of this, it is difficult to avoid exposure to *Toxoplasma gondii* infection (Weiss and Kim 2004).

4.7. Toxoplasmosis Prevalence in Humans

Toxoplasmosis is one of the most common parasitic zoonoses in the world. In 1939, Sabin (cit. by Tenter et al. 2000) demonstrated for the first time that *Toxoplasma* isolated from humans and *Toxoplasma* from animals belong to the same species (Tenter et al. 2000). In 1948, the introduction of the methylene blue test by Sabin and Feldman made it possible to develop seroepidemiological studies in humans and a large number of animal species. Since then, it is estimated that about one-third of the world's population has been exposed to infection with this parasite, but the prevalence of infection in humans is very different between different countries, between different geographical areas of a country and between different ethnic groups in the same area (dos Santos et al. 2010). In the last three decades, variations in the presence of *T. gondii* antibodies have been obtained from 0 to 100% in different adult human populations (Macpherson 2005; Dubey and Jones 2008).

When comparing disease seroprevalence data, it should be taken into account that the different methods used to obtain these data are not standardized. The Sabin–Feldman colorimetric test, which is considered to be "golden" in detecting antibodies in humans, is laborious and has a disadvantage, as it requires a continuous stock of live parasites. For the detection of antibodies, there are several types of serological tests (DAT, ELISA, IFAT, IHAT, LAT, SFDT) that differ in terms of sensitivity, specificity and resulting values (Cilievici and Crețu 2007; Dubey et al. 2014a).

Two tests do not give the same results even if they are performed in the same laboratory. In addition, the prevalence rate differs over time and depending on the age of the individuals studied (Barabas-Hajdu et al. 2007). All of these tests compare the prevalence of the disease based on age, human culture, environmental factors or other factors that may influence the epidemiology of the disease. For example, in the 1990s (Dubey and Lappin 1998; Cook et al. 2000; Tenter et al. 2000), in some Central European countries such as Austria, Belgium, France, Germany and Switzerland, a prevalence of between 37 and 58% was estimated in pregnant women with no obstetric history. A similar seroprevalence was observed in Croatia, Poland, Slovenia, Australia and North Africa. In contrast, an increased seroprevalence was observed in some Latin American countries such as Argentina, Brazil, Cuba, Jamaica and Venezuela, between 51 and 72%, and in West African countries between 54 and 77%. A lower seroprevalence was reported in South Asia, China and Korea (4–39%) and in cold climates, such as the Scandinavian countries (11–28%) (Lindová et al. 2006). Messier et al. (2009) identified a prevalence of 59.8% in Inuit in Quebec, Canada (Messier et al. 2009). Low prevalence of toxoplasmosis has also been found in vegetarians (Hall et al. 1999; Roghmann et al. 1999).

4.7.1. *Toxoplasma gondii* Infection in Human Adults

While *T. gondii* infection is very common in humans, the clinical expression of the disease is more common in at-risk groups. In immunocompetent people, the disease most often develops asymptomatically. Occasionally, mild symptoms may occur, of which lymphadenopathy is the most significant clinical manifestation. Encephalitis, septicemic syndrome, myocarditis or hepatitis may also occur, but these are very rare in immunocompetent people (Dubey 2005).

Since the early 1950s, *T. gondii* infection has been recognized as an important cause of chorioretinitis, with ocular toxoplasmosis being considered the result of a prenatal infection that manifests later in life (Soiza and Pastor 1970). While chorioretinitis lesions in congenital toxoplasmosis are well known, the question has arisen as to whether similar ocular lesions in older children or adults are the result of a recent infection, primary infection or reactivation of prenatal infection.

While previous studies have relied on sporadic cases of *Toxoplasma* chorioretinitis, recent studies (Radbea et al. 2006a) have examined numerous cases due to the outbreak of acute toxoplasmosis in adults due to various causes. In cases where the source of infection was established by epidemiological research, the period between the primary infection and the appearance of clinical signs was between one month and 3.5 years, and the age of patients ranged from 10 to 57 years, but most patients develop the lesions retinally less than one year after infection (Dubey and Beattie 1990; Fan et al. 2007).

In immunocompromised people, a previous latent infection may reactivate, with manifestations of encephalitis. *Toxoplasma* encephalitis and disseminated toxoplasmosis have been observed in patients with immunosuppression due to various causes. Disseminated toxoplasmosis can also complicate organ or bone marrow transplants, or it can occur as a result of transplants from a donor infected with *T. gondii* to a susceptible receptor, or by reactivating a latent receptor infection, due to immunosuppressive treatments (Durecko et al. 2004).

T. gondii is an opportunistic pathogenic factor in people with AIDS. Worldwide, *T. gondii* causes severe encephalitis in over 40% of AIDS patients, and 10–30% of them die from the disease (Erscoiu et al. 2007). By restoring immunity and specific therapy, the incidence of CNS toxoplasmosis in AIDS patients is now declining in many countries, and the reactivation of a latent infection can be prevented by prophylactic treatment with trimethroprim–sulfamethoxazole (Erscoiu et al. 2007).

4.7.2. Congenital Toxoplasmosis

In 1923, Janku et al., (cit. Tenter et al. 2000) first described *Toxoplasma gondii* tissue cysts in the retina of an 11-month-old child with hydrocephalus and congenital microphthalmia (Tenter et al. 2000). This was later recognized as the first case of congenital toxoplasmosis in a human. In the late 1930s, the studies of Wolf and Cowen, (cit. by Tenter et al. 2000) demonstrated that *T. gondii* is the causative agent of encephalomyelitis in newborns (Tenter et al. 2000).

Data on the description of toxoplasmosis in humans also come from 1938, in America. *T. gondii* was identified in a newborn girl who, at the age of 3 days, had seizures and macular lesions in both eyes. The baby died one month after birth. Following necropsy, extra- and intracellular *T. gondii* tachyzoites from encephalomyelitis and retinitis lesions were identified (Figures 16–19). Samples were taken from the cerebral cortex and spinal cord, which were homogenized with saline and inoculated intracerebrally in rabbits and mice. The animals developed encephalitis, the parasite could be isolated from neural lesions, and the passage on other mice resulted in success (Desmonts and Couvreur 1974).

Clinical manifestations in toxoplasmosis:

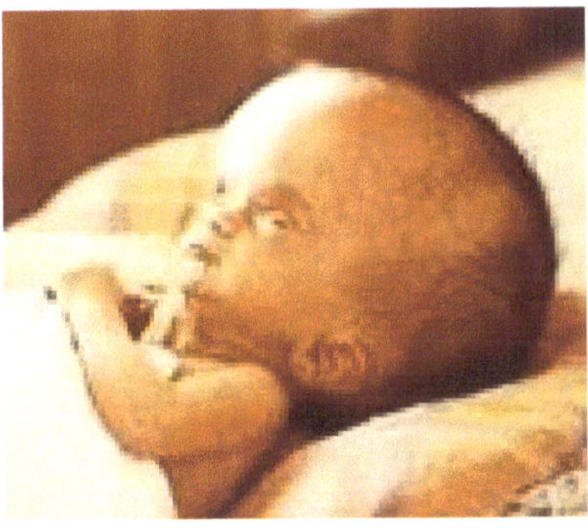

Figure 16. Signs of hydrocephalus (Congenital Toxoplasmosis). Source: Available online: www.medical-labs.net/toxoplasma-gondii-2986/(accessed on 06 March 2019).

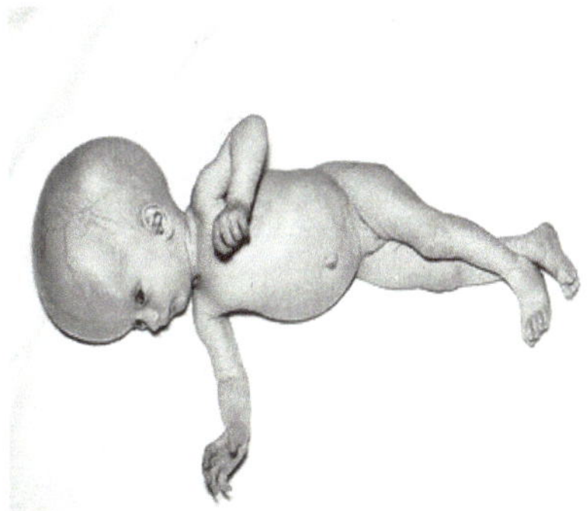

Figure 17. Skull increased and brain decreased due to fluid accumulation (Neonatal hydrocephalus associated with toxoplasmosis). Source: Available online: www.ma ternofetal.net/en/toxoplasmosis-the-pregnant-woman-and-the-cats/(accessed on 19 October 2018).

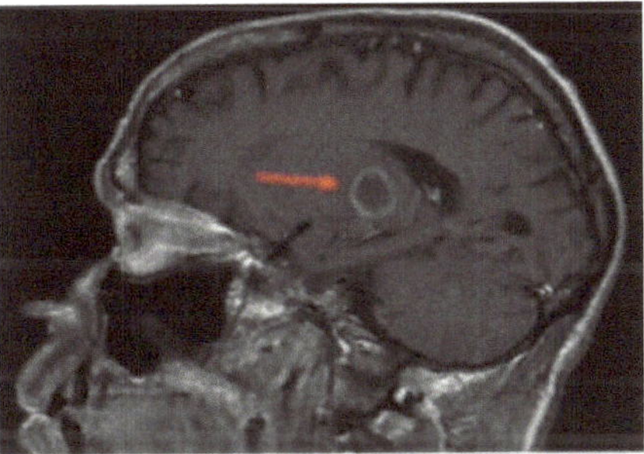

Figure 18. Intracerebral calcification (arrow) (*Toxoplasma* periventricular calcifica-tion). Source: Available online: www.medical-labs.net/toxoplasma-gondii-298 6/(accessed on 6 March 2019).

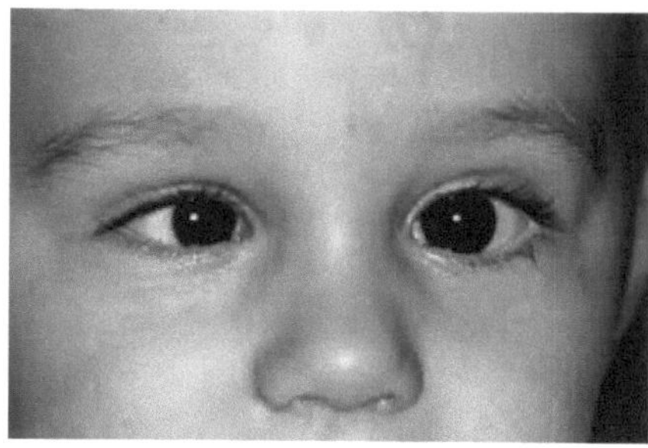

Figure 19. Strabismus and visual disorders. Source: Reprinted from (Fahnehjelm et al. 2000).

In immunocompetent individuals, if the primary infection occurs 4–6 months before conception, or even earlier, the developed immunity will prevent vertical transmission to the fetus in case of subsequent exposures. Exceptions are immuno-compromised women with systemic lupus erythematosus or immunodeficiency syndrome (AIDS) in whom previous infections are transmitted congenitally (Dunn et al. 1999; Erscoiu et al. 2007).

If the first infection occurs during pregnancy, *T. gondii* is transmitted to the fetus even in immunocompetent women. In the last weeks of pregnancy, the transmission rate is over 90%, and in the first 6 weeks, the rate is below 2%. The average transmission rate is 14% in the first trimester, 30% in the second trimester and 59% in the third trimester of pregnancy (Teodorescu et al. 2007).

The mechanism of vertical transmission is still unknown (MacLaren et al. 2004). It seems that temporary parasitemia would be important in the first infection of pregnant women, which could cause the invasion of the placenta by tachyzoites. Eventually, some tachyzoites could cross the placenta and thus reach the fetal circulation or fetal tissues. Congenital toxoplasmosis can lead to miscarriage, neonatal death or malformations in the fetus, as well as a lower quality of life in fetuses that survive (Tenter et al. 2000; Olariu et al. 2008a, 2008b). The risk of intrauterine infection of the fetus, the risk of congenital toxoplasmosis and the severity of the disease depend on the time of infection during pregnancy, the number and virulence of parasites transmitted to the fetus and the age of the fetus at the time of transmission (Wang et al. 2006). If not treated in time, the risk of intrauterine infection of the fetus increases during pregnancy from about 14% after the first infection of the mother in the first trimester to about 59% after the first infection of the mother in the last trimester.

The incidence also varies depending on the methods of estimating the disease. Estimation can be made directly, considering the number of survivors at birth or in the immediate aftermath, or indirectly, considering children who may survive *T. gondii* infection during pregnancy. Recent estimates are based on serological studies, suggesting an incidence of primary infection during pregnancy from 1 to 310 out of 10,000 pregnancies. These percentages are slightly higher if only susceptible women who did not develop immunity before conception are taken into account (Tenter et al. 2000). If the risk of intrauterine infection of the fetus is increased, then the effects are all the more severe, as the transmission takes place in an earlier stage of pregnancy. The most significant manifestation is encephalomyelitis, which can have severe consequences (Neagoe et al. 2007).

About 10% of prenatal infections end in abortion or neonatal death, and another 10–23% of prenatal infections result in clinical signs in newborns from birth. The classic signs of congenital toxoplasmosis—chorioretinitis, intracranial calcifications and hydrocephalus—occur in about 10% of these newborns, while in other newborns, there are a variety of symptoms, from nervous symptoms to non-specific symptoms of an acute infection: convulsions, splenomegaly, hepatomegaly, fever, anemia, lymphadenopathy, etc. About 12–16% of these newborns die (Radbea et al. 2006b). Survivors have mental retardation or other neurological impairments that require special education. If the infection occurs in a more advanced stage of pregnancy, the effects on the fetus are less severe, and those who are infected in the last trimester of pregnancy are asymptomatic at birth. Approximately 67–80% of prenatal infections are subclinical and can only be diagnosed using serological or other laboratory methods (Neagoe et al. 2007). These newborns appear healthy at birth, but they may have clinical symptoms or various deficiencies later in life. These deficiencies mainly affect the eyes (chorioretinitis, strabismus, blindness), the CNS (psychomotor or other neurological deficiencies) or the hearing (deafness). It has been estimated that in about one-third of prenatal infections, children will have visual problems later in life (Olariu et al. 2008b).

4.7.3. Behavioral Disorders Induced by *Toxoplasma gondii* Infection

In order to study whether toxoplasmosis induces various behavioral changes in humans, laboratory tests were performed and questionnaires were applied. Thus, it has been observed that people with chronic toxoplasmosis have an attention deficit (Havlícek et al. 2001). By serologically examining people involved in various car accidents, it was found that many individuals had anti-*Toxoplasma* antibodies. The researchers concluded that people with chronic toxoplasmosis are twice as likely to cause car accidents as people who are seronegative (Flegr et al. 2002, 2009; Yereli et al. 2006; Kocazeybek et al. 2009).

Congenital toxoplasmosis, contracted in the first trimester of pregnancy, causes intracranial calcifications, mental retardation, deafness, seizures and retinal damage in children. If the infection occurs in the last trimester of pregnancy, the child will have a lower IQ and other sequelae that may occur later in life (Alford et al. 1974; Wilson et al. 1980). In congenital toxoplasmosis, psychic sequelae (psychosis or schizophrenia) could not be demonstrated. However, in postnatal toxoplasmosis, and in the chronic phase of the disease, more and more studies confirm the association of mental illness with this parasitic infection (Kramer 1966; Ladee 1966). An example of this is the case of a 20-year-old man who had delirium, auditory hallucinations and catatonic manifestations and who was diagnosed with *Toxoplasma* encephalitis (Freytag and Haas 1979). Other authors have associated suicidal tendencies with elevated serum anti-*Toxoplasma* antibody titers (Yagmur et al. 2010).

Some studies suggest that different strains of *T. gondii* cause different behavioral changes in experimentally infested mice (Kannan et al. 2010). Type I is known to be more lethal to mice than types II and III, but it is not known whether this may be true for humans. Starting with this premise, it was observed that type I is more frequently involved in the production of psychotic symptoms, especially affective psychoses, in people with chronic toxoplasmosis, compared to types II and III (Xiao et al. 2009).

It has been known for more than 40 years that neurotransmitters are involved in the pathogenesis of schizophrenia. The excess of dopamine is the most incriminating factor involved in this disease. Researcher Henry Stibbs (1985) began the study of toxoplasmosis, stimulated by the idea of the effects produced by *T. gondii* on learning and attention, memory and behavior in mice and rats. He infected mice with the C56 strain of *T. gondii*, which he sacrificed at different stages of the disease. Some were sacrificed in the acute phase of toxoplasmosis, and some in the chronic phase. The brains were evaluated neurochemically, and it was observed that in the acute phase of the disease, the level of dopamine was not altered compared to normal limits, but in the chronic phase, this level was increased by 14%. The study concluded that *T. gondii* causes changes in the metabolism of catecholamines. This could contribute to the psychological and motor changes encountered in experimentally infested mice (Stibbs 1985). Other researchers suggest that elevated dopamine levels are caused by activated cytokines due to the presence of the *T. gondii* parasite in the body (Flegr et al. 2003; Flegr 2007).

In 2009, Dr. Glenn McConkey of the University of Leeds in the United Kingdom demonstrated that *T. gondii* has genes that encode two of the enzymes needed to make dopamine. The parasite has genes for phenylalanine hydroxylase, which converts phenylalanine to tyrosine, and genes for tyrosine hydroxylase, which converts tyrosine to dopa, the precursor of dopamine. These genes could not be found in other parasites except for Neospora. This research suggests the possibility that, in individuals with schizophrenia and chronic toxoplasmosis, the excess dopamine identified may be induced by the parasite (Gaskell et al. 2009). Antipsychotic medication with dopamine inhibitory action is used to treat schizophrenia. Administration of haloperidol to mice experimentally infested with *T. gondii* resulted in an improvement in toxoplasmosis-induced behavior, explained by the ability of haloperidol to inhibit parasitic replication and thus implicitly reduce dopamine levels (Webster et al. 2006; Webster and McConkey 2010).

5. Methods of Diagnosis in Toxoplasmosis

5.1. Epidemiological Diagnosis

Epidemiological diagnosis considers the possible cohabitation of farm animals with cats and the possibility of them defecating in feed depots. In humans, the disease can be suspected if the patient eats poorly cooked meat or if he is a cat owner (Dărăbuş et al. 2006). In the case of parasitized animals, the epidemiological diagnosis may be suspected depending on their living environment and maintenance (rural, yard, contact with other animal species) and their diet (raw meat, game). In sheep, transmission can be achieved through grains that are administered in the last part of pregnancy, but also through the consumption of infected placentas, in pigs through ear and tail biting and in cats through the consumption of game species (Tenter et al. 2000).

5.2. Clinical Diagnosis

Establishing a definite clinical diagnosis is impossible due to the non-specific symptoms. In intermediate hosts, the symptoms are similar, with differences in the intensity of symptoms depending on the susceptibility of the species to this disease. In cats, parasitism with *T. gondii* develops asymptomatically.

Clinical signs are more severe in transplacental transmission. Animals infected with *T. gondii* can abort or give birth to non-viable products, which are much smaller than normal. Some fetuses may be of normal size but may have fatal clinical manifestations from the first days of life to the age of 1 year: diarrhea, dyspnea, cough, hyperthermia, ataxia, weight loss and mortality in a proportion of 50%. Those who survive are usually left with nervous (ataxia) and ocular (visual disturbances) sequelae (Dubey 2005).

Postnatal infested animals show clinical signs that may be the cause of pulmonary, neuro-muscular localizations or the cause of generalized infection. The neurological form of the disease can last for several weeks without affecting other systems, while severe damage to the lungs or liver can cause animal death (Klein et al. 2010). Myocardial damage is usually subclinical; arrhythmias and other heart disorders rarely occur (Dărăbuş et al. 2006). Neurological signs also depend on the location of the lesion: at the cerebral, cerebellar or spinal level and can be observed nervous breakdowns, cranial nerve damage, tremor, ataxia, paresis or paralysis. Animals with myositis first have abnormal muscle dysfunction, and then paraparesis or tetraparesis can progress rapidly to paralysis of motor neurons (Cosoroabă 2005).

Also, anterior or posterior uveitis in one or both eyes is common, and iritis, iridocyclitis, or chorioretinitis may develop on its own or at the same time. Crystalline dislocation, glaucoma or retinal displacement are common manifestations of uveitis, and chorioretinitis may occur in the area of tapetum lucidum or tapetum nigricans. Ocular toxoplasmosis occurs in animals with polysystemic clinical signs (Dubey 2005).

A statistical study found that in 100 cats with histologically confirmed toxoplasmosis, the clinical signs were diverse, but lung infections (97.7%), CNS infections (96.4%), liver infections (93.3%), cardiac (86.4%), pancreatic (84.4%) and ocular (81.5%)

infections were the most common (Dubey and Lappin 1998). In experimental *Toxo-plasma* infestation, cats infected with IVF (feline immunodeficiency virus) had severe forms of pneumonia and hepatitis, while cats not infected with IVF had multifocal chorioretinitis and previous uveitis (Tenter et al. 2000).

Behavioral changes: Pigs infected with *T. gondii* appear to show more cannibalism actions (Macpherson 2005; Lindová et al. 2006; Wang et al. 2006).

As early as the late 1970s, researchers Piekarski and Witting began studying behavioral changes induced by chronic *T. gondii* infection. Early observations suggested decreased learning ability in mice and rats and impaired memory in mice. Other researchers (Hutchinson and Hay) have observed hyperactivity, especially in exploring a new environment (Holliman 1997). Continuing research, Webster (1994) observed that experimentally infested mice were more active and were not frightened by the smell of cat urine (Berdoy et al. 2000). Due to the lack of conservation instinct for danger, mice can be hunted more easily by cats (Webster 2001, 2010). Moreover, Vyas et al. (2007) suggests that mice are not only not frightened by the smell of cat urine, but even have an attraction to the pheromones released by the final host (Vyas et al. 2007). These behavioral changes ensure the completion of the biological cycle of the *T. gondii* parasite and may represent a hypothesis of "manipulation" by the parasite on the host. The results of research suggest that *T. gondii* induces specific brain changes. These hypotheses are also supported by Gulinello et al.'s (2010) studies which show that experimentally infested mice have a complex brain pathology as well as motor and sensory deficits, but cognitive abilities are not impaired (Gulinello et al. 2010).

5.3. Diagnosis in Slaughterhouse

The necropsy diagnosis at the slaughterhouse is very difficult to make. Focal necrotic lesions of a few mm, located in the muscles, lungs, liver, spleen and, possibly, in the nerve centers, must attract attention (Nesbakken, Truls 2009. Food Safety in a Global Market—Do We Need to Worry?). However, it is very difficult to impose a detailed necropsy examination of each carcass, even if it is known to come from HIV-positive animals (Jay et al. 2005).

In the absence of lesions, parasites can be detected after peptic digestion, by histological sections stained by silver impregnation or by lamellae display of the brain substance. But these examinations cannot be practiced systematically in slaughterhouses (Colville and Berryhill 2007).

Serological diagnosis is possible in slaughter animals, but is neither practical nor economically justified (Sickinger et al. 2009).

5.4. Morphopathological Diagnosis

The anatomopathological diagnosis mainly concerns the presence of necrotic foci in various tissues and organs (brain, lungs, liver, mesenteric lymph nodes). In intermediate host species, macro- and microscopic lesions are similar to those in cats. There are pneumonias, encephalitis, keratitis, corneal ulcers and abortions (Dărăbuş et al. 2006).

Typical miliary necrosis lesions can be seen on fetal coatings and abortions. Some fetuses can be macerated or mummified. Abortions present with intestinal congestion, cerebral congestion, infiltrated connective tissue, serosanguinous fluids in the abdominal, thoracic and pericardial cavities (Tenter et al. 2000). In full-term infants, but suffering from a congenital toxoplasmosis, various malformations can be found (hydrocephalus, hypotrophy, paraplegia) (Dubey and Lappin 1998).

At the pulmonary level, whitish-gray nodules, approximately 5 mm, can be seen, located subpleural and in the parenchyma and fibrinous exudate. Bronchial lymph nodes are enlarged and necrotic. Necrotic foci have also been identified in the pancreas, liver, kidneys and spleen (Dărăbuş et al. 2006). Multiple ulcers, about 10 mm, were identified in the stomach and small intestine. Cholangiohepatitis occurs only in cats, not in other intermediate host species (Tenter et al. 2000).

Discolorations and necrosis, approximately 12 mm in diameter, and cerebral atrophy were observed in the CNS. At the neural level, necrosis, gliosis and vasculitis are characteristic of non-suppurative meningoencephalomyelitis in foci. Myositis can occur in the muscles of the limbs, but also in the myocardium. Affected muscles are pale, reduced in volume, and in severe or chronic cases are replaced by connective tissue (Zhang et al. 2009).

5.5. Histopathological Diagnosis

Tachyzoites can be identified in various tissues and fluids of the body by histological examination during the acute phase of the disease. They are rarely found in the blood (parasitemia lasting only 1–2 days), CSF, in aspiration or washing fluids, tracheal or bronchoalveolar, but are often found in thoracic or peritoneal fluids, or in thoracic effusion or ascites fluids (Petersen et al. 2006). The presence of tachyzoites in histological sections or in organic fluids, demonstrated by the method of immunoperoxidase with monoclonal antibodies, establishes a definite diagnosis. This technique has been shown to be very useful in the diagnosis of CNS lesions in immunosuppressed individuals (Palm et al. 2008). The microscopic diagnosis is made by examining smears of fetal or gastrointestinal fluids or fingerprints of the brain, spleen, lungs, liver, heart, cotyledons, stained May-Grunwald-Giemsa (M.G.G). For histopathological examination, sections of the brain that are stained with Giemsa can be performed (Palm et al. 2008).

In dead animals, following the acute evolution of toxoplasmosis, foci of inflammation with mononuclear infiltrate can be identified, with or without necrosis in the foci, in a series of tissues and organs, such as liver, heart and lungs (Dubey and Lappin 1998). Necrosis and inflammation are usually associated with the presence of tachyzoites.

Affected placental cotyledons show foci of coagulation necrosis, in case of abortion or fetal death in sheep and goats, these areas mineralizing over time (Buxton et al. 2007). In some cases, a small number of intra- and extracellular parasites can be observed at the periphery of necrotic areas or in newly infested villas (Dubey and Lappin 1998).

Primary and secondary lesions may develop in the fetal brain. Microglial foci with the center usually necrotic and sometimes mineralized are associated with lymphoid meningitis in low-intensity foci and represent an immune response of the fetus in relation to the lesions caused by the local multiplication of the parasite. *Toxoplasma* cysts are rarely identified, usually at the periphery of these lesions. Outbreaks of leukomalacia are another common pathological change and are considered a result of fetal anoxia caused by placental lesions, which prevent adequate transfer of oxygen from mother to fetus (Pereira-Bueno et al. 2004). Such foci are usually identified in cerebral white matter and less frequently in cerebellar matter. Confirmation of the identity of parasite-like structures in tissue sections can be achieved by immunohistochemical technique that stains intact parasites and antigenic debris. Other suitable methods are indirect ABC immunoperoxidase and antiperoxidase (PAP) technique.[1]

Cytological and histopathological investigations do not allow the detection of *T. gondii* tissue cysts in muscle tissue samples, because the tissue density of *Toxoplasma* cysts can be lower than 1/50 g of tissue, which drastically limits the possibility of identification by histological means (Garcia et al. 2006a).

Histopathological, by staining hematoxylin-eosin preparations, tissue cysts appear in the form of circular structures, 5–50 μm, filled with crescent-shaped bradyzoites, colored in blue.

An alternative method of evaluating nerve tissue is to take a small portion of the rostral segment of the brain (dimensions approximate to those of a matchstick), which will then be compressed between the blade and the blade. Tissue cysts are easily detected on microscopic examination.

The impossibility of detecting tissue cysts does not exclude a positive diagnosis. In this case, murine sera are evaluated for the presence of anti-*Toxoplasma* antibodies (IFAT). *Toxoplasma* infection is not considered if the result of this evaluation is also negative (see Footnote 1).

5.6. Experimental Diagnosis

The presence of *T. gondii* can be confirmed by i.p. inoculation, on laboratory animals or on cell cultures. Laboratory mice are the most susceptible animals. Isolation of the parasite can be done from the brain of aborted fetuses and fetal membranes, from secretions, excretions, pathological fluids and other tissues, which we suspect to be infected with *T. gondii*, obtained at necropsy or biopsy. Oocysts obtained from feces and spores can also be used for this purpose (Gatkowska et al. 2006). Tissue samples should not be frozen because the process kills parasitic elements (Zhang et al. 2009).

For bioprobes, *T. gondii*-free mice are used, in which 1×10^7 tachyzoites from the RH strain of *T. gondii* are inoculated intraperitoneally (harvested immediately before from other mice or tissue cultures). Mice will be euthanized, their peritoneal fluid will be harvested and mixed with an equal volume of sterile PBS (Garcia et al. 2006a). Starting with day 4–6 p.i. peritoneal exudate of mice that have been inoculated i.p. is examined for the presence of *T. gondii* tachyzoites. Tissue cysts, more commonly located in neural tissue, are observed from week 4–6 p.i., and antibodies appear from

---

[1] OIE Terrestrial Manual, Chap. 1.09.10. Toxoplasmosis. Available online: https://www.oie.int/en/?s=&_search=Toxoplasma (accessed on 5 February 2020).

week 6 p.i. The optimal time to collect tachyzoites is 72 h after the initial infection, during which time they are found in a sufficiently large number (Cosoroabă 2005). At 7–14 days ascites appears, with the possibility of highlighting the parasite. This method is also used for the differential diagnosis of Hammondia hammondi, in which case myositis occurs at about 30 days (Krasteva et al. 2009).

5.7. Radiological Diagnosis

Radiographic images of the thorax, especially in animals in the acute phase of the disease, reveal the appearance of a diffuse interstitial alveolar aspect that goes until the appearance of the model with lobar distribution in foci (Dubey and Lappin 1998). Symmetrical, diffuse and homogeneous increase in thoracic density due to the destruction of the alveolar walls and the union of the pulmonary alveoli was found in animals with severe forms of the disease. A moderate pleural effusion may also be present (Lago et al. 2007).

At the abdominal level, radiographs reveal the presence of opaque masses in the intestines and mesenteric lymph nodes, or a homogeneous increase in radiographic density as a result of effusion. Loss of radiographic contrast in the right abdominal quarter may indicate the presence of pancreatitis (Dubey 2005).

5.8. Aqueous Humor and Cerebrospinal Fluid Examination

In animals with encephalitis or *Toxoplasma* uveitis, proteins and leukocytes may be increased in CSF or aqueous humor. Small or large mononuclear cells and neutrophils can be found, but the parasitic organism is rarely observed in these fluids (Hendrix and Sirois 2007).

When specific antibodies are observed in these fluids, local production must first be differentiated from passive diffusion of antibodies due to vascular barrier destruction. In the case of serological tests for locations other than the eye or the CNS, a comparison can be made of antibodies in the aqueous humor and CSF with those in the blood, as well as against other agents of other infections that may have the same locations.

This coefficient can be calculated as follows in Equation (1) (Dubey 2005):

$$Antibody\ Coeff. = \frac{Anti-Toxoplasma\ antibodies\ from\ the\ watery\ humor\ (or\ CSF)}{Anti-Toxoplasma\ antibodies\ from\ the\ blood} \times \frac{Specific\ antibodies\ to\ other\ agents\ from\ the\ blood}{Specific\ antibodies\ to\ other\ agents\ from\ the\ watery\ humor\ (or\ CSF)} \tag{1}$$

Titers or equivalent numerical values obtained by ELISA or other methods will be replaced in this formula (Ghazaei 2006). Values of antibody coefficient higher than 1 and especially higher than 8 are considered much more concrete evidence of local antibody production and associated to *Toxoplasma* infestations, respectively, than of the presence of antibodies due to vascular destruction (Dubey 2005).

Occasional local production of anti-*Toxoplasma* antibodies in the eye (Ig G and Ig A) or in the CNS (Ig G) has been observed and studied after experimental infections. However, local Ig M production detected only in naturally infested animals with uveitis or clinical signs of encephalitis suggests that this class of CSF Ig or aqueous humor may be markers of clinical toxoplasmosis (Dubey and Beattie 1990; Ghazaei 2006).

Toxoplasma in the aqueous humor of infested animals can be detected by PCR. Usually, the parasite is identified before it detects the production of antibodies in the eye. Because *T. gondii* can be identified from aqueous humor even in the absence of signs of uveitis, a positive result by PCR does not certify that the ocular signs are of *Toxoplasma* etiology (Contini et al. 2005).

In animals of economic interest, diagnostic methods have not been so thoroughly investigated, probably due to the economic destination of these species and their short life. However, ocular changes, similar to those in humans, can sometimes be observed clinically: keratitis, corneal ulcers, hypopyon accompanied by periorbital edema and lagophthalmia (McAllister 2005; Cilievici and Creţu 2007).

5.9. Haematological Examination

Routine haematological parameters may be altered in animals with acute systemic toxoplasmosis. These parameters have been studied more intensively in cats than in animals of economic interest. The most common changes are: anemia, neutrophilic leukocytosis, lymphocytosis, monocytosis and eosinophilia (Dubey 2005).

Leukopenia, found in some cats with toxoplasmosis, can persist until death and is usually characterized by lymphopenia and absolute neutropenia, eosinopenia and monocytopenia (Calderaro et al. 2009). In experimentally infected cats, neutropenia and lymphopenia may persist for 5–12 days. Leukocytosis was observed in the remission phase of the disease. Lymphocyte counts of more than 7,000 cells/μl were observed between the 28th and 154th day after the first inoculation. The second inoculation did not produce significant changes in the number of leukocytes (Dubey and Lappin 1998).

5.10. Biochemical Examination

Biochemical changes during the acute phase of the disease included hypoproteinemia and hypoalbuminemia, and hyperglobulinemia was observed in animals with chronic toxoplasmosis. Increases in serum alanine aminotransferase (ALT) and aspartate aminotransferase (AST) have been observed in animals with acute hepatic and muscular necrosis. Serum creatine kinase activity is also increased in muscular necrosis (Dubey 2005).

Serum bilirubin levels have also been increased in animals with hepatic necrosis, especially in those with cholangiohepatitis or hepatic lipidosis (Dubey and Lappin 1998). In animals with pancreatitis, there was an increase in serum amylase and lipase, as well as a decrease in total serum Ca, with a normal albuminemia concentration. Animals may often have proteinuria and bilirubinuria (Dubey and Beattie 1990; Calderaro et al. 2009).

5.11.1. Serological Examination

Several serological tests are currently available for the detection of anti- *T. gondii* antibodies (Lindsay and Weiss 2004).

The dye test (DT), described by Sabin and Feldman in 1948, was the first diagnostic method applied to identify specific anti-*Toxoplasma* antibodies. The color test is considered the "gold standard" of serological tests for the detection of anti-*Toxoplasma* antibodies. Live *T. gondii* tachyzoites are incubated for one hour at 37 ° C with a complement-like accessory factor and test serum, then methylene blue solution is added. Specific antibodies induce permeabilization of the tachyzoite membrane and, consequently, release of the cytoplasm from them. For this reason, tachyzoites incubated in the presence of positive sera (with anti- *T. gondii* antibodies) appear uncolored. On the contrary, the incubation of tachyzoites in the presence of negative sera (which do not contain specific antibodies) and the application of methylene blue lead to their staining in blue. DT is specific and sensitive in human patients, but may be inaccurate in other species. In addition, it is very risky because it involves the use of live tachyzoites. It should be noted that, in view of the new animal welfare requirements, the maintenance of tachyzoites in the peritoneum of laboratory mice should be avoided, using them whenever possible in cell cultures (Montoya 2002).

Indirect immunofluorescence for antibodies—IFAT was introduced in practice in the 1960s. It is a simple and widely used technique, specific, but with lower sensitivity than the color reaction. Whole and inactivated *Toxoplasma gondii* tachyzoites are incubated with the diluted test serum, then the appropriate anti-species fluorescent serum is added, and the result is visualized under a fluorescence microscope. In the case of negative sera, tachyzoites will appear in red due to the autofluorescence of the blue Evan dye, and may also have a cap with green fluorescence at the pole (nonspecific polar fluorescence). Tachyzoites will show red fluorescence on at least 80% of their surface and will be surrounded by an uninterrupted band of green fluorescence in the presence of positive sera (Talari et al. 2004).

The direct agglutination reaction—DAT is sensitive and specific. Formalized *Toxoplasma gondii* tachyzoites are introduced into U-shaped microtiter plates, the serum to be analyzed being applied over them. The positive samples will cause agglutination that can be graded, and the negative ones the precipitation of tachyzoites at the base of the wells. The test is easy, but requires a large number of parasitic antigens. DAT and latex agglutination (LAT) methods are not species specific, which is why they can be applied to all species. Indirect hemagglutination (IHA) does not require active antigens, but is less sensitive than the staining test or IFA. Latex agglutination (LAT) has a slightly better sensitivity, but cannot be used to differentiate classes of immunoglobulins (Gardner et al. 2000).

The modified agglutination test (MAT) can only detect Ig G, but is very sensitive compared to other tests. The MAT method has been improved in terms of sensitivity and specificity to detect acute or chronic toxoplasmosis using acetone- or formalin-fixed trophozoites. Antibodies to acetone-fixed antigens are only relevant during acute infestations (less than three months after infestation), while antibodies to formalin-fixed antigens may remain at high titers for several years (Calderaro et al. 2009).

The enzyme-linked immunosorbent assay (ELISA) became available in 1972. The ELISA can detect Ig G, Ig M and Ig A. The original ELISA technique uses a soluble antigen, prepared from tachyzoites belonging to the RH strain of *T. gondii*, applied to the wells of microtiter plates. To this is added the test serum (eg, serum of ovine origin) and then an anti-species enzyme—labeled conjugate, such as horseradish peroxidase—labeled anti-ovine IgG. The attachment of the conjugate causes color changes in the substrate, in direct relation to the number of bound antibodies, which can be read with the aid of a spectrophotometer at the specific absorbance of the substrate used. ELISA is as sensitive as IFA and more sensitive than LAT or IHA (Gross et al. 1993; Glynn 2008).

A kinetic ELISA (KELA) technique has recently been developed. The KELA system measures the rate of reaction between the bound enzyme and the substrate solution that leads to color. The optical density is read three times at 45-s intervals (using the KELA data management program), and the results are expressed as graphs. The correlation between ELISA and KELA results is very high, both being excellent diagnostic tools (Golkar et al. 2008).

Another diagnostic method is the anti-*Toxoplasma* IgG maturation test to determine the time of infection. This test shows an increase in the avidity of specific anti-*Toxoplasma* antibodies in the IgG class as the immune response matures, and the time of infection can be determined within a maximum of three months after infestation (Auer et al. 2000). The test has been adapted to automated systems, and new variants allow the exclusion of acute toxoplasmosis for an interval of 3–4 months ago in patients who possess antibodies with high avidity (Remington et al. 2004). However, the presence of IgG with reduced or intermediate avidity is not equivalent to the diagnosis of the acute form because it is known that the maturation of antibodies shows significant individual differences. In addition, some individuals possess IgG with reduced avidity for several months after infection, and therapy and/or pregnancy may delay IgG maturation (Robert-Gangneux et al. 1999). By comparing the maturation of the IgG response in pregnant and treated women, with that observed in untreated patients and without physiological gestational status, in the latter category a much faster maturation of IgG was observed in the first four months after infection (Roberts et al. 2001). The average value of the avidity index is 0.3 in patients with acute infection, an index that can be kept low (0.1–0.6) in the first nine months after infection (Cozon et al. 1998). The test designed for sheep in order to detect IgG avidity for *T. gondii* P30 antigen is an appropriate diagnostic tool by differentiating relatively recent infections from those with long evolution (Pietkiewicz et al. 2007).

For the detection of acute infection, the association between a sensitive test for the identification of anti-*Toxoplasma* IgM and the determination of the avidity index of anti-*Toxoplasma* IgG possesses the highest predictive value regarding the time of infection (Olariu et al. 2006). However, the Ig G avidity test does not contribute to the diagnosis of reactivated toxoplasmosis (Lachaud et al. 2009; Tekkesin et al. 2011).

Once infected, the animals carry tissue cysts for the rest of their lives. Ig G in chicks born to mothers with chronic infection, is transmitted through colostrum and persists 8–12 weeks after birth. Serological controls indicate that *T. gondii* infection is widespread worldwide (Wallace 1969; Chinchilla et al. 1994; Roghmann et al. 1999). The prevalence of HIV-positive cases increases with the age of the animals due to the high possibilities of exposure (Cosoroabă 2005).

The biggest problem of the diagnostic act is the situation in which specific anti-*Toxoplasma* antibodies are identified in the first sample collected after conception, because the moment of infection becomes a key element for estimating the degree of risk in which the fetus is (Pinon et al. 2001). The response to anti-*Toxoplasma* IgG antibodies develops in the first eight weeks after infection, after which the IgG titer is maintained at a high level, which may or may not be accompanied by a decline in IgM antibodies. Low titers of anti-*Toxoplasma* IgM can be detected up to several years after an acute infection, which demonstrates that low levels of anti-*Toxoplasma* antibodies are not a sign of acute infection (Petithory et al. 1996). Individuals positive for anti-*Toxoplasma* IgM, but with a low titer, should be retested every two weeks. Only the association between a technique with high sensitivity in IgM detection with the IgG avidity test ensures excellent diagnostic performance (Siloşi et al. 2007). In contrast to these, methods for detecting IgA or IgE have been shown to be less useful in diagnosing acute infection (Dubey and Lappin 1998).

The establishment of an antemortem diagnosis of a clinical toxoplasmosis should be based on the corroboration of the following parameters:

1. serological evidence of a recent or active infestation, confirmed by high IgM titers, as well as by establishing IgG titers;

2. the exclusion of other causes that could give similar symptoms;

3. clinical positive response to one of the specific treatments for toxoplasmosis (Table 21).

In cats, the positivity of a serological reaction does not provide any information regarding the elimination of oocysts in the feces.

In case of infection of a cat by sporulated oocysts, whose sporozoites must undergo an exenteral evolution before locating as a *Coccidia* in enterocytes, the antibodies (Ig G, Ig M, Ig A) are detectable before starting the elimination of oocysts: Ig M at seven days, Ig A at eight days, Ig G at 15 days and oocysts at 21 days (Talari et al. 2004).

Table 21. Paraclinical elements in *Toxoplasma* gondii infection (by Weiss and Kim 2004).

Infection Type	Paraclinical Signs
Acute infection (e.g., lymphadenitis):	- the serological test for IgM is positive; serial tests will demonstrate an increase in IgG titer.
Acute toxoplasmosis during pregnancy:	- the serological response is similar to that seen in other patients with acute infection; the most important aspect of this situation is the time of infection because IgM can persist for several months after an acute form of toxoplasmosis, and the risk of vertical transmission is associated only with the infection produced during pregnancy. The application of the IgG avidity test (with low index in acute infection), differential agglutination and the detection of the presence of IgA and/or IgE (which appear earlier than IgM) facilitates the definition of the moment of infection and the need to evaluate congenital transmission in utero.
Chronic infection:	- IgM absent, IgG present, but without changes in titer when analyzing serial samples.
Reactivation of the disease in case of immunosuppression (for example, brain toxoplasmosis in HIV/AIDS patients):	- absence of IgM and presence of IgG. Rare situations are also identified in which the serological examination is negative for both classes of antibodies, being necessary to take tissue samples to clarify the diagnosis. Due to the fact that the PCR technique is extremely sensitive in the mentioned situation, a positive result of it has diagnostic value.
Congenital toxoplasmosis:	for the diagnosis in utero, the association of PCR in amniotic fluid and ultrasonography is used. Serological examination of newborns will be positive for IgG (due to maternal antibodies). The presence of IgM or IgA in the newborn confirms the diagnosis of congenital toxoplasmosis, as well as the stable or serously increasing IgG titer.

If the infection is achieved by bradyzoite cysts, after a shorter prepatent period, oocysts appear in the feces faster than antibodies in the blood: on the 4–5th day. In this case, of bradyzoite cyst infestation, Ig A antibodies appear after Ig G and

Ig M (34th week) and are not detectable during the fecal oocyst elimination period (Lachaud et al. 2009).

In all cases, the elimination of oocysts is very short, while the persistence of antibodies is longer: 3–5 months for Ig M or even up to 6 months. Kittens born from a *Toxoplasma* infected mother eliminate oocysts and their blood has IgM, signs of an active infection (Cosoroabă 2005).

Regarding the assessment of the risk to public health, the results of serological tests of cats can be interpreted as follows:

-   a seronegative cat is not eliminating oocysts, but can become exposed to some sources. This category of cats poses the greatest risk to public health;
-   a seropositive cat may not eliminate oocysts, especially in the case of reinfestation or immunosuppression (Remington et al. 2004).

Due to the fact that antibodies can appear in the blood of both healthy and sick cats, serological tests alone cannot reveal a clinical toxoplasmosis. Because IgM class antibodies are most often found in the aqueous humor of clinically or IVF-infected cats, but not in healthy cats, it has been concluded that IgM are more eloquent markers for clinical toxoplasmosis. than IgG and IgA (Thiele 1990). A negative serological response suggests that the cat has not yet been exposed to *T. gondii* infection and is still susceptible to infection in the future (Cosoroabă 2005).

5.11.2. Oral Fluids Examination

More and more studies highlight a new diagnostic method for the detection of Ig A and IgG anti-*Toxoplasma* antibodies in oral fluids. This method is non-invasive, applicable to patients of any age, including newborns. Comparative studies of oral fluids and peripheral blood, performed comparatively on the same groups of patients, highlight the increased relevance of this method in the rapid diagnosis of toxoplasmosis (Stroehle et al. 2005; Sampaio et al. 2014; Chapey et al. 2015).

In humans, the use of oral fluids to detect Ig G antibodies is increasingly specified in specialized articles. Using various methods, such as Western Blot, inhouse ELISA test, or others, subclasses of antibodies specific for *T. gondii* infection can be identified. The presence of antibody subclasses in oral fluid may be the body's first line of defense. Studies on this topic have shown a difference in the profile of IgG antibody subclasses between saliva and serum. Thus, it has been shown that there is a richer immune response in saliva, as several subclasses of antibodies coexist at that level, unlike serum. These results highlight the major role of the oral immune system, which may have the ability to limit systemic *T. gondii* infection, an area of study that requires further research (Chapey et al. 2015; Styles et al. 2018).

More recently, such studies have also been performed in experimental *T. gondii* infections in animals. Campero et al. 2020, and Teixeira et al. 2020, tested the effectiveness of oral fluid analysis to determine anti-*Toxoplasma* antibodies in animals (pigs and cats). The researchers showed that antibodies against *T. gondii* can be detected in oral fluids from experimentally infected animals and that IgA in particular was more relevant compared to IgG. These results can be used as a basis for the development of additional diagnostic tests, with a higher yield and a lower cost. This inexpensive, non-invasive method, maintaining animal welfare, could be a screening option for animals, even on a farm, to determine exposure to *T. gondii* infection (Campero et al. 2020; Teixeira et al. 2020).

5.12. Molecular Techniques of Diagnosis—Polymerase Chain Reaction (PCR)

Molecular biology is based on techniques that allow the detection or capture of tiny amounts of nucleic acids. With the introduction of the polymerization chain reaction—PCR, higher levels of detection sensitivity and amplification of specific DNA sequences were achieved in a very short time. PCR is a relatively simple method, requiring a very small amount of biological material, whereby a DNA template or cDNA is amplified quickly and safely, thousands or millions of times[2].

Several PCR techniques have now been developed for the detection of *T. gondii* DNA. The main target regions are the repetitive sequences of the B1 and P30 genes (SAG1) and the 18S ribosomal RNA (rRNA). The sensitivity of the chain polymerization reactions is dependent on the number of copies of the target sequences (P30: one copy; B1: 35 copies; rRNA: 110 repeating units) (see Footnote 2). The B1 repetitive sequence amplification method has been used to evaluate lens aspirates in congenitally infected human patients and has been shown to be more sensitive than the conventional diagnostic method (ELISA) (Li et al. 2006; Béla et al. 2008). Although PCR is a very sensitive technique, the interpretation of its results should be done with caution if applied as the only diagnostic method. A definite diagnosis can be made following the association of PCR with other diagnostic methods (Thiele 1990; Petersen et al. 2001).

Real-time PCR technique allows simultaneous quantification and amplification of DNA (Kasper et al. 2009). The method is similar to other PCR variants and can be performed in 96-well microtiter plates (Harris and Jones 1997). The fluorogenic dye is intercalated with double-stranded DNA after each amplification step, and the results, illustrated on the amplification graph, allow the quantification of the parasitic elements in the DNA sample. The method was used to amplify and quantify the DNA of the B1 gene of *T. gondii*. Quantification of parasitic DNA can be used to determine the number of parasites in tissue or fluid samples, such as amniotic fluid in patients suspected of congenital infection (Cermáková et al. 2005; Su et al. 2006).

The next method described is the nested form of PCR, which amplifies the repetitive DNA sequence of the B1 gene (Kocher et al. 1990). Parasitic DNA can be extracted and purified from several types of tissue, including the placenta, central nervous system, heart, and skeletal muscle. Erythrocytes are removed from tissue samples by lysis buffer washing and centrifugation. The DNA is extracted and then proteinase K is inactivated by boiling. B1 gene amplification is performed with two sense primers P1 and P2, two antisense primers P3 and P4 and 2.5 units of Taq polymerase (White 1996).

Acute toxoplasmosis can be diagnosed by detecting anti-*Toxoplasma* antibodies using serological tests or *T. gondii* DNA in organic fluids or tissue samples. Identifi-

---

2 PCR ACCESS_ PCR Protocols and Reference Guide—Promega. Available online: https://www.prom ega.ro/products/pcr/rt-pcr/access-rt-pcr-system/?catNum=A1260 (accessed on 15 March 2020).

cation of tachyzoite DNA in body fluids and tissue samples using PCR amplification is an effective method of diagnosing congenital and cerebral toxoplasmosis (Weiss and Kim 2004). Diagnostic techniques based on the detection of parasitic DNA use biological fluids (amniotic fluid and/or fetal blood, CSF) and tissues as samples (Fuentes et al. 1996; Nagy et al. 2006). Inoculation of these biological samples into mice or fibroblast cultures are frequently used techniques to establish a definite diagnosis. The mouse test allows the identification of viable parasites and the PCR of the parasitic DNA, even if the tissues are in the decomposition phase (Weiss 1995). PCR cannot highlight the risk of human infection because it does not distinguish between viable and non-viable parasitic elements. However, the human infectious dose is less than 104 tachyzoites. Thus, the sensitivity of the PCR technique for the 529 bp fragment (10–102 tachyzoites/ml) should be sufficient to establish the risk situation of consumers (Garcia et al. 2006b).

The disseminated form of toxoplasmosis can affect the brain, lungs and heart. An association of serological results, in vitro culture and clinical examination is required to establish the diagnosis. The analysis of bronchoalveolar fluid obtained by lavage using PCR with primers for genes B1 and P30 is much more sensitive than microscopy in the detection of parasitic elements of *T. gondii* (Petersen et al. 2006).

PCR testing for B1 and P30 genes for blood samples for parasitemia is a high-potential technique, all the more so as it uses biological samples taken by non-invasive methods to diagnose disseminated toxoplasmosis. PCR has been shown to be more sensitive than monitoring for therapeutic effects in suspected cerebral toxoplasmosis. The level of parasitism can be approximated based on the intensity of the PCR signal by analyzing the agarose gel stained with ethidium bromide (Weiss 1995).

There are also PCR methods for co-diagnosis in toxoplasmosis (Isaac-Renton et al. 1998). Considering that microscopic examination of feces is not sensitive enough to observe *T. gondii* oocysts, and inoculation on laboratory animals is long and expensive, PCR methods have been adapted to determine the DNA of T. oocysts. fecal gondii (Dumètre and Dardé 2007; Dumètre et al. 2008).

Before use, the feces are subjected to mechanical vortexing and repeated freeze-thaw cycles in order to destroy the wall of the sporulated oocysts. The fecal samples are then prepared by the double centrifugation method, but with several centrifugation cycles to concentrate the oocysts. Samples are stored at $-20\,^\circ$C before being processed by PCR for DNA extraction (Dumètre and Dardé 2005; Sotiriadou and Karanis 2008).

Thus, Salant et al. (2007) tested the coprodiagnostic PCR method in toxoplasmosis. Out of 122 fecal samples from cats in Jerusalem, 11 positive results were found, while microscopic examination did not show any positive samples (Salant et al. 2007).

It should be noted that cats begin to eliminate oocysts before antibodies form in the blood. PCR methods thus become very important in establishing an early diagnosis and the acute phase of elimination of *T. gondii* oocysts (Harris and Jones 1997).

Given that infestation of humans and animals is often achieved through contaminated water or soil, these methods can also be adapted for the determination of

T. gondii oocysts in water and soil samples (Aramini et al. 1999; Afonso et al. 2008; Lass et al. 2009; Schares et al. 2008).

5.13. Coprological Diagnosis in Cats

Despite the high prevalence of antibodies in the serum of cats worldwide, the prevalence of *T. gondii* oocysts in their feces is low. In the United States, less than 1% of cats are found to eliminate oocysts (Dubey and Jones 2008). Because cats usually eliminate oocysts only 1–2 days after the primary infection, routine coproscopic examinations rarely detect *T. gondii* oocysts (Airinei et al. 2007). Moreover, cats usually show no clinical signs of disease and do not have diarrhea during the elimination of oocysts. Although cats are considered refractory to a reocclusion of oocysts, they may eliminate a few oocysts after a new infection with another strain of *T. gondii* if more than 6 years have passed since the last infection. In the intestine of immune cats, only a part of the asexual development of the parasite takes place, compared to cats that have never had an infection, in which a complete developmental cycle takes place. In cats with chronic infection, immunosuppression with high doses of prednisolone (10–18 mg/kg orally, daily, or i.m., weekly) will result in a reuptake of oocysts, while a lower dose mg/kg i.m., for 4 weeks) will not have this effect (Garcia et al. 2007).

In cat feces, oocysts of *T. gondii* cannot be morphologically differentiated from oocysts of *Hammondia hammondi* or *Besnoitia darlingi*. Oocysts of these coccidia can be differentiated only by sporulation or by inoculation on animals (Schares et al. 2008). *T. gondii* oocysts are one-fourth the size of *Isospora felis* oocysts and one-eighth the size of *Toxocara cati* eggs (which may be most common in cats) (Dubey 2009a).

Classical fecal examination methods are fully suitable for the detection of protozoan oocysts. The technique of direct smears, with sodium chloride solution and staining, becomes the optimal procedure for examining a fecal sample in order to detect protozoa.

Staining can be used to identify certain structural characteristics of trophozoites and oocysts. Lugol and methylene blue solutions are most used for direct smear staining. These solutions do not preserve the blade, but facilitate the examination of the specimen, making identification easier (Ferguson 2009).

Fecal flotation methods are based on the specific gravity of the parasitic material and fecal debris. Specific gravity refers to the weight of an object in relation to an equal volume of water. The specific gravity of parasite eggs is 1100–1200 g/ml, while the specific gravity of water is 1000 (Hendrix and Sirois 2007). In order for protozoan oocysts to float, the solution used must have a higher specific gravity than that of the parasitic material. Some saline or carbohydrate solutions are very suitable for flotation. Most have a specific gravity of 1200–1250. In general, heavier fecal debris descends to the base of the solution container and the parasitic material rises to its surface (Colville and Berryhill 2007).

The sodium nitrate solution (315 g of sodium nitrate, 1 l of water) is the variant most frequently used in veterinary clinics, allowing the flotation of oocysts. Sodium nitrate can form crystals and, in case of prolonged exposure, can deform eggs (Montoya 2002).

Another commonly used solution is carbohydrate. It does not involve high costs, does not crystallize and does not deform the eggs. Due to their very small size, *T. gondii* oocysts are best highlighted by centrifugation using Sheather sugar solution. First, 5–10 grams of feces are mixed with water until an aqueous consistency is obtained, and then the mixture is strained through gauze. Two parts of Sheather sugar solution (500 g of sugar, 300 ml of water and 6.5 g of melted phenol crystals) are added over one part of the fecal suspension, placed in a coated centrifuge tube and then centrifuged. After centrifugation at 1000 (rotations/minute) for 10 min, 1–2 drops are taken with a pipette, placed on a slide, covered with a slide and examined under a microscope with the objective of 100 (Dubey 2005).

The zinc sulphate solution (386 g of zinc sulphate, 1 l of water) is usually used in diagnostic laboratories. Zinc sulfate induces protozoan flotation and a minimum degree of deformation. It is usually used in combination with one of the solutions mentioned above (Zhang et al. 2009).

The last convenient option is the saturated sodium chloride solution (salt is added to the water until it no longer dissolves and tends to settle to the bottom of the container). However, it is corrosive to laboratory equipment, forms crystals and significantly deforms parasite eggs. It is also a weak environment for flotation because its maximum specific gravity is 1200 (it allows heavier eggs to settle) (Hendrix and Sirois 2007).

Standard or simple fecal flotation is the most commonly applied technique in veterinary clinics. The method involves the use of a test tube or ampoule in which the feces and flotation solution are mixed. The tube will be covered with a microscope slide and left to rest. Fecal eggs that reach the surface of the solution will adhere to the microscope slide. After removal from the tube, the slide will be examined under a microscope for parasitic material. Although the standard flotation technique is extremely easy, it does not have the same efficiency as the centrifugation method (Calderaro et al. 2009). Commercial flotation kits include filter ampoules and flotation solution.

Highlighting oocysts via centrifugal flotation is the best method available. A concentration technique allows the examination of a large amount of feces in a relatively short time, requiring less time than the standard technique. The only disadvantage of this method is that it requires a centrifuge with a rotor adapted for 15 mL tubes (fecal samples mixed with flotation solution are centrifuged for 3–5 min at 1300–1500 rpm). If such a device is available, we consider that centrifugal flotation is preferable because the method is easy and the samples can be processed both individually and in batches (Dubey 2006).

The double centrifugation technique involves centrifuging the fecal samples diluted with water at 1500 rpm for 10 min. This is followed by the second centrifugation, also at 1500 rpm, for 10 min, in which the supernatant initially obtained is replaced with carbohydrate solution (Dubey 2005).

Quantitative procedures aim to determine the number of eggs or oocysts present in one gram of feces, which is considered a general indicator of parasitic load. However, their usefulness is limited by the fact that, in case of polyparasitism, several types of parasitic material are present. Additionally, the elimination of oocysts may be sporadic, and their number may not be correlated with the actual parasitic load (Hendrix and Sirois 2007).

The diagnosis of toxoplasmosis in cats is necessary in measures to prevent transmission to women, especially pregnant women or individuals with compromised immunity. In such contexts, examination of feces is not the best procedure, as many cats deposit most oocysts in just 1–2 days, while the entire patent period can take up to 20 days. These days, only a few oocysts are removed and exams can be negative, although there are eliminations of oocysts (Cosoroabă 2005; Krasteva et al. 2009).

6. Elements of Epidemiology and Recommendations for Reducing Toxoplasmosis Seroprevalence

Information collected in a study in Romania allows the synthesis of identified epidemiological elements that contribute to maintaining a high prevalence of toxoplasmosis, and the enunciation of some recommendations (Verdeș 2011).

Thus, the transmission of the disease from cat to cat, through oocysts, makes it possible to resume the biological cycle of *Toxoplasma gondii* with the possibility of environmental contamination. The transmission of the disease from cats to farm animals is important both due to the possibility of infestation of other intermediate hosts, via carnivorism, and the risk of resuming the biological cycle of the parasite if that contaminated meat is eaten by parasitological negative cats. Additionally, the main danger is the possibility of transmitting the disease to humans or, moreover, to pregnant women. Due to the cats' habit of burying their excrement, feed depots are a suitable place for defecation. In this way, a large number of animals can be contaminated even if they are kept away from cats. The high prevalence obtained in this study highlights the risk of transmitting the disease to humans not only through contact with oocysts eliminated by cats, but also through the consumption of insufficiently cooked infected meat. This depends on the degree of civilization, culture and culinary habits of the people in the area. Therefore, removing cats from the house is not a solution to prevent the disease.

Based on the epidemiological investigations carried out and the results obtained, in order to reduce the prevalence of toxoplasmosis, the following measures are recommended:

In animals (Figures 20 and 22):

- Feeding cats only with commercial food and not with raw meat, organs or bones;
- If, however, meat (or food scraps) is administered in animal feed, it must be cooked;
- Prohibiting cats from going outside the house (apartment), thus preventing them from hunting;
- Destruction or control of the population of rodents, kitchen beetles, etc., on farms and in households;
- Bins must always be covered to prevent the consumption of household waste, corpses, etc.;
- Cats near farms must be neutered to keep the feline population under control;
- The presence of cats near pregnant animals should not be allowed;
- The construction of spaces specially arranged for the storage of fodder, to which the cats do not have access and the animal feed is well protected from atmospheric factors;
- The movement of cats in animal shelters or in rooms where animal feed is processed shall be prohibited;
- Feed quality-control;
- Improving hygiene conditions in animal shelters;

- No household waste will be administered in the feed of pigs without being sterilized beforehand (by boiling);
- Dead animals must be isolated immediately to prevent cannibalism in pigs and necrophagy in pigs;
- Fetal membranes and abortions must be incinerated to prevent contamination of cats and other farm animals;
- Preventing the entry of birds into animal shelters;
- Observance of hygiene by farm care staff.
- In humans (Figure 21):
- Rigorous personal hygiene;
- Washing hands with soap and water after stroking cats;
- Prohibiting the cat from accessing the bed, the table or the kitchen shelves;
- Rigorous washing of fruits and vegetables;
- Non-consumption of water from rivers and springs, if it has not been filtered;
- Wearing gloves for gardening and agriculture;
- Daily change of sand in cat litter;
- Hand washing after changing the cat's sand, or wearing gloves;
- Pregnant women are not to change the cat's sand; if choosing to do so, they must wear gloves and a mask during this operation;
- Avoiding very close contact with pets;
- Consumption of well-cooked meat (cooked at over 70 °C in the center of the piece of meat, or frozen at—20 °C);
- Washing surfaces and kitchen utensils with water and detergent;
- The tasting of raw meat (especially by pregnant women) is prohibited.

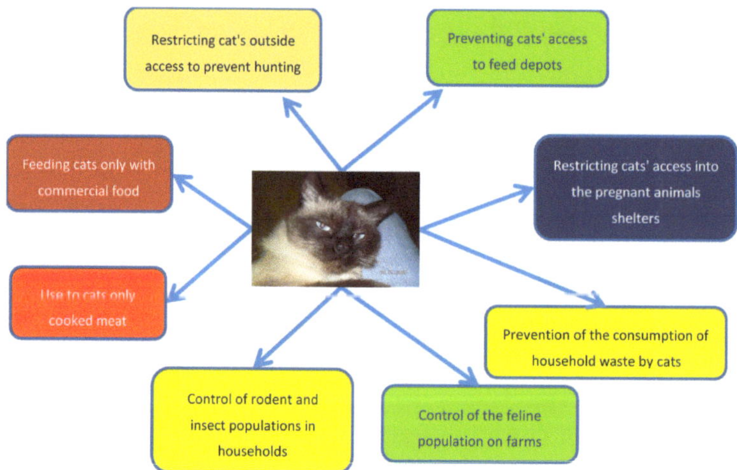

Figure 20. Preventive measures for the definitive host. Source: Figure by authors.

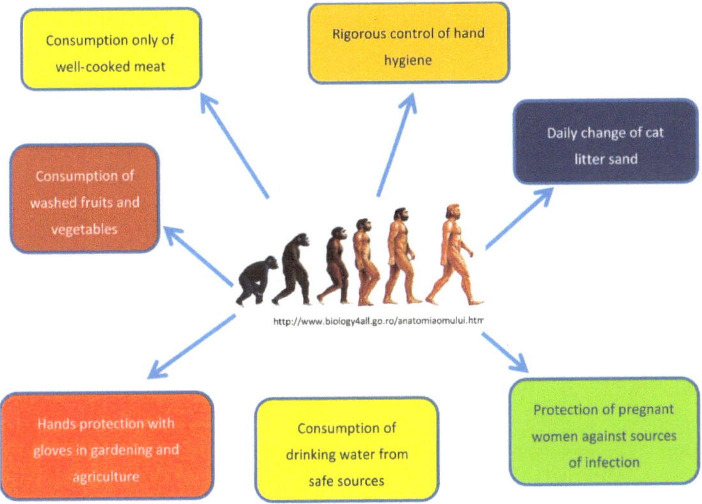

Figure 21. Preventive measures for humans. Source: Figure by authors.

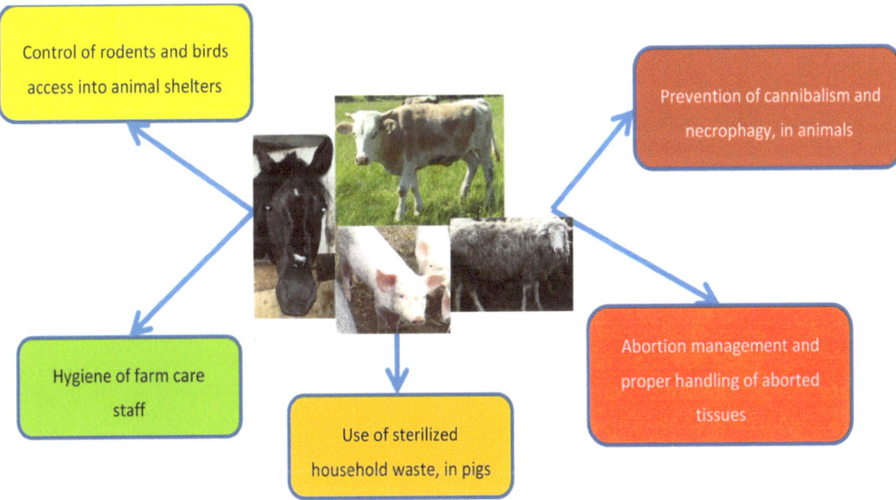

Figure 22. Preventive measures for the intermediate host. Source: Figure by authors.

The prophylaxis measures described above can significantly reduce the risk of *T. gondii* infection but cannot completely prevent it.

References

Afonso, Eve, Estelle Germain, Marie-Lazarine Poulle, Sandrine Ruette, Sébastien Devillard, Ludovic Say, Isabelle Villena, Dominique Aubert, and Emmanuelle Gilot-Fromont. 2013. Environmental Determinants of Spatial and Temporal Variations in the Transmission of *Toxoplasma gondii* in Its Definitive Hosts. *International Journal for Parasitology. Parasites and Wildlife* 2: 278–85. [CrossRef]

Afonso, Eve, Mélissa Lemoine, Marie-Lazarine Poulle, Marie-Caroline Ravat, Stéphane Romand, Philippe Thulliez, and Isabelle Villena. 2008. Spatial Distribution of Soil Contamination by *Toxoplasma gondii* in Relation to Cat Defecation Behaviour in an Urban Area. *International Journal for Parasitology* 38: 1017–23. [CrossRef] [PubMed]

Ahmad, Nisar, H. Ahmed, Shamaila Irum, and Mazhar Qayyum. 2014. Seroprevalence of IgG and IgM Antibodies and Associated Risk Factors for Toxoplasmosis in Cats and Dogs from Sub-Tropical Arid Parts of Pakistan. *Tropical Biomedicine* 31: 777–84. [PubMed]

Aigner, Cleiton Paulo, Aristeu Vieira da Silva, Fabiano Sandrini, Paulo de Sá Osório, Lilian Poiares, and Alvaro Largura. 2010. Real-Time PCR-Based Quantification of *Toxoplasma gondii* in Tissue Samples of Serologically Positive Outdoor Chickens. *Memorias Do Instituto Oswaldo Cruz* 105: 935–37. [CrossRef] [PubMed]

Airinei, Nicolae, Gheorghe Mertoiu, and Nicolae Ciucă. 2007. Frecvenţa Toxoplasmozei La Câinii Şi Pisicile de Companie, Examinate Coproparazitar. *Revista Română de Parazitologie* 17: 103.

Ajioka, James W., Jennifer M. Fitzpatrick, and Christoper P. Reitter. 2001. "*Toxoplasma gondii* genomics: shedding light on pathogenesis and chemotherapy." *Expert Reviews in Molecular Medicine* 3(1):1-19. doi:10.1017/S1462399401002204

Akca, Atila, Cahit Babur, Mukremin Ozkan Arslan, Yunus Gicik, Murat Kara, and S. Kilic. 2023. Prevalence of Antibodies to *Toxoplasma gondii* in Horses in the Province of Kars, Turkey. *Veterinarni Medicina—UZPI (Czech Republic).* Available online: https://scholar.google.com/scholar_lookup?q=Prevalence+of+antibodies+to+Toxoplasma+gondii+in+horses+in+the+province+of+Kars%2C+Turkey (accessed on 18 March 2023).

Akhtardanesh, Baharak, Naser Ziaali, Hamid Sharifi, and Shirin Rezaei. 2010. Feline Immunodeficiency Virus, Feline Leukemia Virus and *Toxoplasma gondii* in Stray and Household Cats in Kerman-Iran: Seroprevalence and Correlation with Clinical and Laboratory Findings. *Research in Veterinary Science* 89: 306–10. [CrossRef] [PubMed]

Al Hamada, Ali, Ihab Habib, Anne Barnes, and Ian Robertson. 2019. Risk Factors Associated with Seropositivity to Toxoplasma among Sheep and Goats in Northern Iraq. *Veterinary Parasitology, Regional Studies and Reports* 15: 100264. [CrossRef] [PubMed]

Albuquerque, George R., Alexandre D. Munhoz, Walter Flausino, Rogério T. Silva, Cláudio R. R. Almeida, Simoni M. Medeiros, and Carlos Wilson G. Lopes. 2005. [Prevalence of anti-Toxoplasma gondii antibodies in dairy cattle from Sul Fluminense Paraíba Valley, State of Rio de Janeiro]. *Revista Brasileira De Parasitologia Veterinaria = Brazilian Journal of Veterinary Parasitology: Orgao Oficial Do Colegio Brasileiro De Parasitologia Veterinaria* 14: 125–28. [PubMed]

Alford, C. A., S. Stagno, and David W. Reynolds. 1974. Congenital Toxoplasmosis: Clinical, Laboratory, and Therapeutic Considerations, with Special Reference to Subclinical Disease. *Bulletin of the New York Academy of Medicine* 50: 160–81. [PubMed]

Ali, C. N., J. A. Harris, J. D. Watkins, and Abiodun Adesiyun. 2003. Seroepidemiology of *Toxoplasma gondii* in Dogs in Trinidad and Tobago. *Veterinary Parasitology* 113: 179–87. [CrossRef]

Allén, Elisa. 2016. *Toxoplasma gondii* -Vasta-Aineiden Esiintyvyys Suomalaisilla Naudoilla. *Lisensiaatintutkielma* 45–47.

Almeida, Daniela, João Quirino, Patricia Fereira Barradas, Priscilla Gomes da Silva, Maria Pereira, Rita Cruz, Carla Santos, Ana Cristina Mega, Fernando Esteves, Carmen Nóbrega, and et al. 2021. A 2-Year Longitudinal Seroepidemiological Evaluation of Toxoplasma gondii Antibodies in a Cohort of Autochthonous Sheep from Central Portugal. *Pathogens* 10: 40. [CrossRef] [PubMed]

Almería, Sonia, Carlos Calvete, A. Pagés, C. Gauss, and Jitender P. Dubey. 2004. Factors Affecting the Seroprevalence of *Toxoplasma gondii* Infection in Wild Rabbits (Oryctolagus Cuniculus) from Spain. *Veterinary Parasitology* 123: 265–70. [CrossRef] [PubMed]

Alvarado-Esquivel, Cosme, C. García-Machado, Domingo Alvarado-Esquivel, J. Vitela-Corrales, Isabelle Villena, and Jitender P. Dubey. 2012a. Seroprevalence of *Toxoplasma gondii* Infection in Domestic Sheep in Durango State, Mexico. *The Journal of Parasitology* 98: 271–73. [CrossRef]

Alvarado-Esquivel, Cosme, M. A. Estrada-Malacón, S. O. Reyes-Hernández, J. A. Pérez-Ramírez, J. I. Trujillo-López, Isabelle Villena, and Jitender P. Dubey. 2012b. High Prevalence of *Toxoplasma gondii* Antibodies in Domestic Pigs in Oaxaca State, Mexico. *The Journal of Parasitology* 98: 1248–50. [CrossRef]

Alvarado-Esquivel, Cosme, S. Rodríguez-Peña, Isabelle Villena, and Jitender P. Dubey. 2012c. Seroprevalence of *Toxoplasma gondii* Infection in Domestic Horses in Durango State, Mexico. *The Journal of Parasitology* 98: 944–45. [CrossRef]

Alvarado-Esquivel, Cosme, Dora Romero-Salas, Anabel Cruz-Romero, Zeferino García-Vázquez, Alvaro Peniche-Cardeña, Nelly Ibarra-Priego, Concepción Ahuja-Aguirre, Adalberto A. Pérez-de-León, and Jitender P. Dubey. 2014. High Prevalence of *Toxoplasma gondii* Antibodies in Dogs in Veracruz, Mexico. *BMC Veterinary Research* 10: 191. [CrossRef] [PubMed]

Alvarado-Esquivel, Cosme, Renata Fabiola Vazquez-Morales, Edgar Eusebio Colado-Romero, Ramiro Guzmán-Sánchez, Oliver Liesenfeld, and Jitender P. Dubey. 2015. Prevalence of Infection with *Toxoplasma gondii* in Landrace and Mixed Breed Pigs Slaughtered in Baja California Sur State, Mexico. *European Journal of Microbiology & Immunology* 5: 112–15. [CrossRef]

Anderson, Seth E., Jr., and Jack S. Remington. 1974. Effect of Normal and Activated Human Macrophages on *Toxoplasma gondii*. *The Journal of Experimental Medicine* 139: 1154–74. [CrossRef] [PubMed]

Antolová, Daniela, Katarína Reiterová, and Pavol Dubinský. 2007. Seroprevalence of *Toxoplasma gondii* in Wild Boars (Sus Scrofa) in the Slovak Republic. *Annals of Agricultural and Environmental Medicine: AAEM* 14: 71–73. [PubMed]

Antoniu, S., B. Ştirbu Teofănescu, D. Militaru, and G. Căpitanu. 2008. Diagnosticul Toxoplasmozei Feline Prin Testele de Imunofluorescenţă Indirectă Şi ELISA. *Revista Română de Medicină Veterinară* 2: 250–54.

Anwar, Ali, Joanna Knaggs, Katrina M. Service, Graeme W. McLaren, Philip Riordan, Chris Newman, Richard J. Delahay, Chris Cheesman, and David W. Macdonald. 2006. Antibodies to *Toxoplasma gondii* in Eurasian Badgers. *Journal of Wildlife Diseases* 42: 179–81. [CrossRef] [PubMed]

Aramini, J. J., C. Stephen, and Jitender P. Dubey. 1998. *Toxoplasma gondii* in Vancouver Island Cougars (Felis Concolor Vancouverensis): Serology and Oocyst Shedding. *The Journal of Parasitology* 84: 438–40. [CrossRef] [PubMed]

Aramini, J. J., C. Stephen, Jitender P. Dubey, Christian Engelstoft, H. Schwantje, and Carl S. Ribble. 1999. Potential Contamination of Drinking Water with *Toxoplasma gondii* Oocysts. *Epidemiology and Infection* 122: 305–15. [CrossRef] [PubMed]

Arko-Mensah, John, K. M. Bosompem, Emmanuel A. Canacoo, Jonathan M. Wastling, and Bartholomew Dicky Akanmori. 2000. The Seroprevalence of Toxoplasmosis in Pigs in Ghana. *Acta Tropica* 76: 27–31. [CrossRef]

Arkush, Kristen D., Melissa A. Miller, Christian M. Leutenegger, Ian A. Gardner, Andrea E. Packham, Anja R. Heckeroth, Astrid M. Tenter, Bradd C. Barr, and Patricia A. Conrad. 2003. Molecular and Bioassay-Based Detection of *Toxoplasma gondii* Oocyst Uptake by Mussels (Mytilus Galloprovincialis). *International Journal for Parasitology* 33: 1087–97. [CrossRef]

Arnaudov, Atanas. 2003. Study on the Toxoplasmosis among Wild Animals. *Experimental Pathology and Parsitology* 6: 51–54.

Aroussi, Abdelkrim, Philippe Vignoles, François Dalmay, Laurence Wimel, Marie-Laure Dardé, Aurélien Mercier, and Daniel Ajzenberg. 2015. Detection of *Toxoplasma gondii* DNA in Horse Meat from Supermarkets in France and Performance Evaluation of Two Serological Tests. *Parasite (Paris, France)* 22: 14. [CrossRef]

Arunvipas, Pipat, Sathaporn Jittapalapong, Tawin Inpankaew, Nongnuch Pinyopanuwat, Wissanuwat Chimnoi, and Soichi Maruyama. 2013. 'Seroprevalence and Risk Factors Influenced Transmission of *Toxoplasma gondii* in Dogs and Cats in Dairy Farms in Western Thailand'. *African Journal of Agricultural Research* 8: 591–95. [CrossRef]

Asgari, Qasem, Bahador Sarkari, Maryam Amerinia, Saed Panahi, Iraj Mohammadpour, and Afrooz Sadeghi Sarvestani. 2013. Toxoplasma Infection in Farm Animals: A Seroepidemiological Survey in Fars Province, South of Iran. *Jundishapur Journal of Microbiology, June.* [CrossRef]

Asgari, Qasem, Iraj Mohammadpour, Razieh Pirzad, Mohsen Kalantari, Mohammad Hossein Motazedian, and Shahrbanou Naderi. 2018. Molecular and Serological Detection of *Toxoplasma gondii* in Stray Cats in Shiraz, South-Central, Iran. *Iranian Journal of Parasitology* 13: 430–39. [PubMed]

Ashmawy, Karam Imam, Somaia Saif Abuakkada, and A. M. Awad. 2011. Seroprevalence of Antibodies to Encephalitozoon Cuniculi and *Toxoplasma gondii* in Farmed Domestic Rabbits in Egypt. *Zoonoses and Public Health* 58: 357–64. [CrossRef]

Aslantaş, Özkan, C. Babur, and Selcuk Kilic. 2001. Seroprevalence of Brucellosis and Toxoplasmosis in Horses in Kars. *Etlik Vet Mikrobiyol. Derg* 12: 1–7.

Aslantaş, Ozkan, Vildan Ozdemir, Selçuk Kiliç, and Cahit Babür. 2005. Seroepidemiology of Leptospirosis, Toxoplasmosis, and Leishmaniosis among Dogs in Ankara, Turkey. *Veterinary Parasitology* 129: 187–91. [CrossRef]

Asthana, Sumita P., Calum N. L. Macpherson, Stanley H. Weiss, Richard Stephens, Thomas N. Denny, R. N. Sharma, and J. P. Dubey. 2006. Seroprevalence of *Toxoplasma gondii* in Pregnant Women and Cats in Grenada, West Indies. *The Journal of Parasitology* 92: 644–45. [CrossRef] [PubMed]

Aubert, Dominique, Daniel Ajzenberg, C. Richomme, Emmanuelle Gilot-Fromont, Marie Eve Terrier, C. de Gevigney, and Yvette Game. 2010. Molecular and Biological Characteristics of *Toxoplasma gondii* Isolates from Wildlife in France. *Veterinary Parasitology* 171: 346–49. [CrossRef] [PubMed]

Auer, Herbert, Angelika Vander-Möse, Otto Picher, Julia Walochnik, and Horst Aspöck. 2000. Clinical and Diagnostic Relevance of the Toxoplasma IgG Avidity Test in the Serological Surveillance of Pregnant Women in Austria. *Parasitology Research* 86: 965–70. [CrossRef] [PubMed]

Aydenizoz, M., Erol Handemir, and K. Kamburgil. 2005. Seroprevalence of Toxoplasmosis in Goats in Kirikkale Province of Turkey. Available online: http://acikerisim.kku.edu.tr/xmlui/handle/20.500.12587/3513 (accessed on 18 March 2023).

Ayinmode, Adekunle B., Daniel O. Oluwayelu, Waidi F. Sule, and Oluwasola O. Obebe. 2014. Seroprevalence of *Toxoplasma gondii* in Recreational Horses in Two Metropolitan Cities of Southwestern Nigeria. *African Journal of Medicine and Medical Sciences* 43: 47–50. [PubMed]

Azevedo, S. S., C. S. A. Batista, S. A. Vasconcellos, D. M. Aguiar, Alessandra Mara Alves Ragozo, A. A. R. Rodrigues, Clebert Jose Alves, and Solange Maria Gennari. 2005. Seroepidemiology of *Toxoplasma gondii* and Neospora Caninum in Dogs from the State of Paraíba, Northeast Region of Brazil. *Research in Veterinary Science* 79: 51–56. [CrossRef] [PubMed]

Balea, Anamaria, Anamaria Ioana Paştiu, Adriana Györke, Viorica Mircean, and Vasile Cozma. 2013. The Dynamics of Anti-*Toxoplasma gondii* Antibodies (IgG) in Small Ruminants and Pigs from Cluj County, Romania. Available online: https://www.semanticscholar.org/paper/The-dynamics-of-anti-Toxoplasma-gondii-antib odies-%28-Balea-Pa%C8%99tiu/9e489e5d39964cd308101c5b63521d3cf6cc056e (accessed on 3 November 2020).

Ballweber, Lora Rickard. 2006. Veterinary Clinics of North America—Food Animal Practice—Ruminant Parasitology. *Veterinary Clinics of North America-Food Animal Practice* 22: 501–712. [CrossRef]

Barabas-Hajdu, Eniko, A. Miklós, T. Miklós, E.M. Bartók, and A. Simó. 2007. Toxoplasmoza Acută În Cazuistica Laboratorului Spitalului Clinic Judeţean de Urgenţă Din Târgu-Mureş În Perioada 2006–2007. *Revista Română de Parazitologie* 17: 105.

Barros, Macarena, Oscar Cabezón, Jitender P. Dubey, Sonia Almería, María P. Ribas, Luis E. Escobar, Barbara Ramos, and Gonzalo Medina-Vogel. 2018. *Toxoplasma gondii* Infection in Wild Mustelids and Cats across an Urban-Rural Gradient. *PLoS ONE* 13: e0199085. [CrossRef] [PubMed]

Bartova, Eva, and Kamil Sedlak. 2012. *Toxoplasma gondii* and Neospora Caninum Antibodies in Goats in the Czech Republic. *Veterinární Medicína* 57: 111–14. [CrossRef]

Bártová, Eva, Kateřina Kobédová, Marie Budíková, and Karol Račka. 2021. Serological and Molecular Detection of *Toxoplasma gondii* in Farm-Reared Ostriches (Struthio Camelus) in the Czech Republic. *International Journal of Food Microbiology* 356: 109333. [CrossRef]

Bártová, Eva, Kamil Sedlák, and Ivan Literák. 2006. Prevalence of *Toxoplasma gondii* and Neospora Caninum Antibodies in Wild Boars in the Czech Republic. *Veterinary Parasitology* 142: 150–53. [CrossRef]

Bártová, Eva, Kamil Sedlák, and Ivan Literák. 2009a. Serologic Survey for Toxoplasmosis in Domestic Birds from the Czech Republic. *Avian Pathology: Journal of the W.V.P.A* 38: 317–20. [CrossRef] [PubMed]

Bártová, Eva, Kamil Sedlák, and Ivan Literák. 2009b. *Toxoplasma gondii* and Neospora Caninum Antibodies in Sheep in the Czech Republic. *Veterinary Parasitology* 161: 131–32. [CrossRef] [PubMed]

Bártová, Eva, Kamil Sedlák, Frantisek Treml, Ivan Holko, and Ivan Literák. 2010. Neospora Caninum and *Toxoplasma gondii* Antibodies in European Brown Hares in the Czech Republic, Slovakia and Austria. *Veterinary Parasitology* 171: 155–58. [CrossRef] [PubMed]

Bártová, Eva, Kamil Sedlák, Kateřina Kobédová, Marie Budíková, Yakubu Joel Atuman, and Joshua Kamani. 2017a. Seroprevalence and Risk Factors of Neospora Spp. and *Toxoplasma gondii* Infections among Horses and Donkeys in Nigeria, West Africa. *Acta Parasitologica* 62: 606–9. [CrossRef] [PubMed]

Bártová, Eva, Kateřina Kobédová, Jiří Lamka, Radim Kotrba, Roman Vodička, and Kamil Sedlák. 2017b. Seroprevalence of Neospora Caninum and *Toxoplasma gondii* in Exotic Ruminants and Camelids in the Czech Republic. *Parasitology Research* 116: 1925–29. [CrossRef]

Bártová, Eva, Radka Lukášová, Roman Vodička, Jiří Váhala, Lukáš Pavlačík, Marie Budíková, and Kamil Sedlák. 2018. Epizootological Study on *Toxoplasma gondii* in Zoo Animals in the Czech Republic. *Acta Tropica* 187: 222–28. [CrossRef] [PubMed]

Basso, Walter, Gastón Moré, Maria Alejandra Quiroga, Lais Pardini, Diana Bacigalupe, Lucila Venturini, M. C. Valenzuela, D. Balducchi, P. Maksimov, G. Schares, and et al. 2009. Isolation and Molecular Characterization of *Toxoplasma gondii* from Captive Slender-Tailed Meerkats (Suricata Suricatta) with Fatal Toxoplasmosis in Argentina. *Veterinary Parasitology* 161: 201–6. [CrossRef] [PubMed]

Basso, Walter, Maria Cecilia Venturini, Gastón Moré, Alejandra Quiroga, Diana Bacigalupe, Juan Manuel Unzaga, Alejandra Larsen, R. Laplace, and Lucila Venturini. 2007. Toxoplasmosis in Captive Bennett's Wallabies (Macropus Rufogriseus) in Argentina. *Veterinary Parasitology* 144: 157–61. [CrossRef]

Basso, Walter, Sonja Hartnack, Lais Pardini, Pavlo Maksimov, Bretislav Koudela, Maria C. Venturini, Gereon Schares, Xaver Sidler, Fraser I. Lewis, and Peter Deplazes. 2013. Assessment of Diagnostic Accuracy of a Commercial ELISA for the Detection of *Toxoplasma gondii* Infection in Pigs Compared with IFAT, TgSAG1-ELISA and Western Blot, Using a Bayesian Latent Class Approach. *International Journal for Parasitology* 43: 565–70. [CrossRef] [PubMed]

Baszler, T. V., Jitender P. Dubey, Christiane V. Löhr, and William J. Foreyt. 2000. Toxoplasmic Encephalitis in a Free-Ranging Rocky Mountain Bighorn Sheep from Washington. *Journal of Wildlife Diseases* 36: 752–54. [CrossRef] [PubMed]

Beck, Josh R., Imilce A. Rodriguez-Fernandez, Jessica Cruz de Leon, My-Hang Huynh, Vern B. Carruthers, Naomi S. Morrissette, and Peter J. Bradley. 2010. A Novel Family of Toxoplasma IMC Proteins Displays a Hierarchical Organization and Functions in Coordinating Parasite Division. *PLoS Pathogens* 6: e1001094. [CrossRef]

Béla, Samantha Ribeiro, Deise A. Oliveira Silva, Jair Pereira Cunha-Júnior, Carlos P. Pirovani, Flávia Andrade Chaves-Borges, Fernando Reis de Carvalho, Taísa Carrijo de Oliveira, and José Roberto Mineo. 2008. Use of SAG2A Recombinant *Toxoplasma gondii* Surface Antigen as a Diagnostic Marker for Human Acute Toxoplasmosis: Analysis of Titers and Avidity of IgG and IgG1 Antibodies. *Diagnostic Microbiology and Infectious Disease* 62: 245–54. [CrossRef]

Belbacha, Imane, Jamaleddine Hafid, R. Tran Manh Sung, P. Flori, H. Raberin, R. Aboufatima, A. Regragui, Anand Dalal, and Abderrahman Chait. 2004. [Toxoplasma gondii: Level of carriage in sheep of Marrakech region (Mnabha)]. *Schweizer Archiv Fur Tierheilkunde* 146: 561–64. [CrossRef]

Belfort-Neto, Rubens, Veronique Nussenblatt, Luiz Rizzo, Cristina Muccioli, Claudio Silveira, Robert Nussenblatt, Assis Khan, L. David Sibley, and Rubens Belfort. 2007. High Prevalence of Unusual Genotypes of *Toxoplasma gondii* Infection in Pork Meat Samples from Erechim, Southern Brazil. *Anais Da Academia Brasileira De Ciencias* 79: 111–14. [CrossRef]

Beltrame, Marcus A. V., Hilda Fatima Jesus Pena, Nielton Cezar Ton, A. J. B. Lino, Solange Maria Gennari, Jitender P. Dubey, and Fausto Edmundo Lima Pereira. 2012. Seroprevalence and Isolation of *Toxoplasma gondii* from Free-Range Chickens from Espírito Santo State, Southeastern Brazil. *Veterinary Parasitology* 188: 225–30. [CrossRef]

Berdoy, Manuel, Joanne P. Webster, and David W. Macdonald. 2000. Fatal Attraction in Rats Infected with *Toxoplasma gondii*. *Proceedings. Biological Sciences* 267: 1591–94. [CrossRef]

Berger-Schoch, A. E., Daland C. Herrmann, Gereon Schares, Norbert Müller, Daniel Bernet, Bruno Gottstein, and Caroline F. Frey. 2011. Prevalence and Genotypes of *Toxoplasma gondii* in Feline Faeces (Oocysts) and Meat from Sheep, Cattle and Pigs in Switzerland. *Veterinary Parasitology* 177: 290–97. [CrossRef] [PubMed]

Bermúdez, Roberto, Luis Daniel Faílde, Ana P. Losada, Jose Maria Nieto, and Maria Isabel Quiroga. 2009. Toxoplasmosis in Bennett's Wallabies (Macropus Rufogriseus) in Spain. *Veterinary Parasitology* 160: 155–58. [CrossRef] [PubMed]

Besné-Mérida, Alejandro, Juan Antonio Figueroa-Castillo, José Juan Martínez-Maya, Héctor Luna-Pastén, Esther Calderón-Segura, and Dolores Correa. 2008. Prevalence of Antibodies against *Toxoplasma gondii* in Domestic Cats from Mexico City. *Veterinary Parasitology* 157: 310–13. [CrossRef] [PubMed]

Bettiol, Silvana S., David Lawrence Obendorf, Mark Nowarkowski, and John Marsden Goldsmid. 2000. Pathology of Experimental Toxoplasmosis in Eastern Barred Bandicoots in Tasmania. *Journal of Wildlife Diseases* 36: 141–44. [CrossRef] [PubMed]

Bier, Nadja Seyhan, Kaya Stollberg, Anne Mayer-Scholl, Annette Johne, Karsten Nöckler, and Martin Richter. 2020. Seroprevalence of *Toxoplasma gondii* in Wild Boar and Deer in Brandenburg, Germany. *Zoonoses and Public Health* 67: 601–6. [CrossRef] [PubMed]

Bisson, A., Stephen Maley, Christopher M. Rubaire-Akiiki, and Jonathan M. Wastling. 2000. The Seroprevalence of Antibodies to *Toxoplasma gondii* in Domestic Goats in Uganda. *Acta Tropica* 76: 33–38. [CrossRef]

Borde, Gustave, G. Lowhar, and Abiodun A. Adesiyun. 2006. *Toxoplasma gondii* and Chlamydophila Abortus in Caprine Abortions in Tobago: A Sero-Epidemiological Study. *Journal of Veterinary Medicine. B, Infectious Diseases and Veterinary Public Health* 53: 188–93. [CrossRef]

Borkovcova, Marie. 2003. Preliminary Report of Toxoplasma/Hammondia Complex, Parasites of Felis Silvestris f. Catus, in Rural Regions of South Moravia (Czech Republic). Acta Universitatis Agriculturae et Silviculturae Mendelianae Brunensis (Czech Republic). Available online: https://scholar.google.com/scholar_lookup?title=Preliminary+report+of+Toxoplasma%2FHammondia+complex%2C+parasites+of+Felis+silvestris+f.+catus%2C+in+rural+regions+of+South+Moravia+%28Czech+Republic%29&author=Borkovcova%2C+M.+%28Mendelova+Zemedelska+a+Lesnicka+Univ.%2C+Brno+%28Czech+Republic%29.+Ustav+Zoologie+a+Vcelarstvi%29&publication_year=2003 (accessed on 15 November 2021).

Bouchard, Émilie, Stacey A. Elmore, Ray T. Alisauskas, Gustaf Samelius, Alvin A. Gajadhar, Keaton Schmidt, Sasha Ross, and Emily J. Jenkins. 2019. Transmission Dynamics of *Toxoplasma gondii* In Arctic Foxes (Vulpes Lagopus): A Long-Term Mark-Recapture Serologic Study at Karrak Lake, Nunavut, Canada. *Journal of Wildlife Diseases* 55: 619–26. [CrossRef]

Boyle, Jon P., and Jay R. Radke. 2009. A History of Studies That Examine the Interactions of Toxoplasma with Its Host Cell: Emphasis on in Vitro Models. *International Journal for Parasitology* 39: 903–14. [CrossRef]

Brandão, Geane Peroni, Adaliene Matos Ferreira, Maria Norma De Melo, and Ricardo Wagner Almeida Vitor. 2006. Characterization of *Toxoplasma gondii* from Domestic Animals from Minas Gerais, Brazil. *Parasite (Paris, France)* 13: 143–49. [CrossRef]

Brennan, Anthea, Jennifer Hawley, Navneet Dhand, Lara Boland, Julia A. Beatty, Michael R. Lappin, and Vanessa R. Barrs. 2020. Seroprevalence and Risk Factors for *Toxoplasma gondii* Infection in Owned Domestic Cats in Australia. *Vector Borne and Zoonotic Diseases (Larchmont, N.Y.)* 20: 275–80. [CrossRef] [PubMed]

Bresciani, Katia Denise Saraiva, A. J. Costa, Gilson Helio Toniollo, M. C. R. Luvizzoto, Cristina Takami Kanamura, Flavio Ruas Moraes, Silvia Helena Venturoli Perri, and Solange Maria Gennari. 2009. Transplacental Transmission of *Toxoplasma gondii* in Reinfected Pregnant Female Canines. *Parasitology Research* 104: 1213–17. [CrossRef] [PubMed]

Bresciani, Katia Denise Saraiva, Solange Maria Gennari, Anna Claudia Marques Serrano, A. a. R. Rodrigues, T. Ueno, L. G. Franco, Silvia Helena Venturoli Perri, and Alessandro F. T. Amarante. 2007. Antibodies to Neospora Caninum and *Toxoplasma gondii* in Domestic Cats from Brazil. *Parasitology Research* 100: 281–85. [CrossRef] [PubMed]

Burns, Roy, Elizabeth S. Williams, Donal O'Toole, and J. P. Dubey. 2003. *Toxoplasma gondii* Infections in Captive Black-Footed Ferrets (Mustela Nigripes), 1992-1998: Clinical Signs, Serology, Pathology, and Prevention. *Journal of Wildlife Diseases* 39: 787–97. [CrossRef]

Buxton, David, Stephen W. Maley, Steve E. Wright, Susan Rodger, Paul Bartley, and Elisabeth A. Innes. 2007. *Toxoplasma gondii* and Ovine Toxoplasmosis: New Aspects of an Old Story. *Veterinary Parasitology* 149: 25–28. [CrossRef]

Cabezón, Oscar, Alisa J. Hall, Cecile Vincent, Marcela Pabón, Ignacio García-Bocanegra, Jitender P. Dubey, and Sonia Almería. 2011. Seroprevalence of *Toxoplasma gondii* in North-Eastern Atlantic Harbor Seal (Phoca Vitulina Vitulina) and Grey Seal (Halichoerus Grypus). *Veterinary Parasitology* 179: 253–56. [CrossRef]

Cabezón, Oscar, Ana R. Resendes, Mariano Domingo, Juan Antonio Raga, Celia Agustí, Ferran Alegre, Jose Luis Mons, Jitender P. Dubey, and Sonia Almería. 2004. Seroprevalence of *Toxoplasma gondii* Antibodies in Wild Dolphins from the Spanish Mediterranean Coast. *The Journal of Parasitology* 90: 643–44. [CrossRef]

Cabezón, Oscar, Javier Millán, Margalida Gomis, Jitender P. Dubey, Ezio Ferroglio, and Sonia Almería. 2010. Kennel Dogs as Sentinels of Leishmania Infantum, *Toxoplasma gondii*, and Neospora Caninum in Majorca Island, Spain. *Parasitology Research* 107: 1505–8. [CrossRef]

Calderaro, Adriana, Simona Peruzzi, Giovanna Piccolo, Chiara Gorrini, Sara Montecchini, S. Rossi, Carlo Chezzi, and Giuseppe Dettori. 2009. Laboratory Diagnosis of *Toxoplasma gondii* Infection. *International Journal of Medical Sciences* 6: 135–36. [CrossRef]

Cambrea, S.C., M. Ilie, M. Cambrea, V. Ionescu, and S. Rugină. 2007. Aspecte Ale Toxoplasmozei Cerebrale La Adolescenţii Seropozitivi. *Revista Română de Medicină Veterinară* 17: 89–93.

Camossi, Lucilene Granuzzio, Haroldo Greca-Júnior, Ana Paula Ferreira Lopes Corrêa, Virginia Bodelao Richini-Pereira, R. C. Silva, Aristeu V. Da Silva, and Helio Langoni. 2011. Detection of *Toxoplasma gondii* DNA in the Milk of Naturally Infected Ewes. *Veterinary Parasitology* 177: 256–61. [CrossRef] [PubMed]

Campero, Lucía María, Franziska Schott, Bruno Gottstein, Peter Deplazes, Xaver Sidler, and Walter Basso. 2020. Detection of Antibodies to *Toxoplasma gondii* in Oral Fluid from Pigs. *International Journal for Parasitology* 50: 349–55. [CrossRef] [PubMed]

Canessa, A., V. Pistoia, S. Roncella, A. Merli, G. Melioli, Alberto Terragna, and M. Ferrarini. 1988. An in Vitro Model for Toxoplasma Infection in Man. Interaction between CD4+ Monoclonal T Cells and Macrophages Results in Killing of Trophozoites. *Journal of Immunology (Baltimore, Md.: 1950)* 140: 3580–88. [CrossRef]

Cano-Terriza, David, María Puig-Ribas, Saúl Jiménez-Ruiz, Óscar Cabezón, Sonia Almería, Ángela Galán-Relaño, Jitender P. Dubey, and Ignacio García-Bocanegra. 2016. Risk Factors of *Toxoplasma gondii* Infection in Hunting, Pet and Watchdogs from Southern Spain and Northern Africa. *Parasitology International* 65: 363–66. [CrossRef]

Cano-Terriza, David, Rafael Guerra, Sylvie Lecollinet, Marta Cerdà-Cuéllar, Oscar Cabezón, Sonia Almería, and Ignacio García-Bocanegra. 2015. Epidemiological Survey of Zoonotic Pathogens in Feral Pigeons (Columba Livia Var. Domestica) and Sympatric Zoo Species in Southern Spain. *Comparative Immunology, Microbiology and Infectious Diseases* 43: 22–27. [CrossRef]

Cañón-Franco, William Alberto, Denise Pimentel Bergamaschi, Leonardo José Richtzenhain, Yeda Nogueira, Luís Marcelo Aranha Camargo, Silvio Luís Pereira de Souza, and Solange Maria Gennari. 2003. Evaluation of the Performance of the Modified Direct Agglutination Test (MAT) for Detection of *Toxoplasma gondii* Antibodies in Dogs. *Brazilian Journal of Veterinary Research and Animal Science* 40: 452–56. [CrossRef]

Caporali, E.H.G., A.V.D.A. Silva, A.D.E.O. Mendonça, and Helio Langoni. 2005. Comparison of Methodologies for Determining the Prevalence of Anti-*Toxoplasma gondii* Antibodies in Pigs from São Paulo and Pernambuco States—Brazil. *Arquivos de Ciências Veterinárias e Zoologia da UNIPAR* 8: 19–24.

Carme, Bernard, Daniel Ajzenberg, Magalie Demar, Stéphane Simon, Marie Laure Dardé, Bertrand Maubert, and Benoît de Thoisy. 2009. Outbreaks of Toxoplasmosis in a Captive Breeding Colony of Squirrel Monkeys. *Veterinary Parasitology* 163: 132–35. [CrossRef]

Carneiro, Ana Carolina Aguiar Vasconcelos, Mariangela Carneiro, Aurora Maria Guimaraes Gouveia, Alessandro S. Guimarães, A. P. R. Marques, L. S. Vilas-Boas, and Ricardo Wagner Almeida Vitor. 2009a. Seroprevalence and Risk Factors of Caprine Toxoplasmosis in Minas Gerais, Brazil. *Veterinary Parasitology* 160: 225–29. [CrossRef]

Carneiro, Ana Carolina Aguiar Vasconcelos, Mariangela Carneiro, Aurora Maria Guimaraes Gouveia, L. S. Vilas-Boas, and Ricardo Wagner Almeida Vitor. 2009b. Seroprevalence and Risk Factors of Sheep Toxoplasmosis in Minas Gerais, Brazil. *Revue de Médecine Vétérinaire* 160: 527–31.

Carrasco, Librado, A. I. Raya, Alejandro Núñez, Jaime Gómez-Laguna, Santiago Hernández, and Jitender P. Dubey. 2006. Fatal Toxoplasmosis and Concurrent Calodium Hepaticum Infection in Korean Squirrels (Tanias Sibericus). *Veterinary Parasitology* 137: 180–83. [CrossRef] [PubMed]

Catenacci, Lilian S., Juliana Griese, Rodrigo C. da Silva, and Helio Langoni. 2010. *Toxoplasma gondii* and *Leishmania* Spp. Infection in Captive Crab-Eating Foxes, Cerdocyon Thous (Carnivora, Canidae) from Brazil. *Veterinary Parasitology* 169: 190–92. [CrossRef] [PubMed]

Cavalcante, Gleicio T., Daniel M. Aguiar, Daniela Chiebao, Jitender P. Dubey, Vera Letticie Azevedo Ruiz, Ricardo Augusto Dias, Luis Marcelo A. Camargo, Marcelo Bahia Labruna, and Solange M. Gennari. 2006. Seroprevalence of *Toxoplasma gondii* Antibodies in Cats and Pigs from Rural Western Amazon, Brazil. *The Journal of Parasitology* 92: 863–64. [CrossRef]

Cazarotto, Chrystian J., Alexandre Balzan, Rhayana K. Grosskopf, Jhonatan P. Boito, Luiza P. Portella, Fernanda F. Vogel, and Juscivete F. Fávero. 2016. Horses Seropositive for *Toxoplasma gondii, Sarcocystis* Spp. and *Neospora* Spp.: Possible Risk Factors for Infection in Brazil. *Microbial Pathogenesis* 99: 30–35. [CrossRef] [PubMed]

Cedillo-Peláez, Carlos, I. D. Díaz-Figueroa, M. I. Jiménez-Seres, G. Sánchez-Hernández, and D. Correa. 2012. Frequency of Antibodies to *Toxoplasma gondii* in Stray Dogs of Oaxaca, México. *The Journal of Parasitology* 98: 871–72. [CrossRef]

Cedillo-Peláez, Carlos, Claudia Patricia Rico-Torres, Carlos Gerardo Salas-Garrido, and Dolores Correa. 2011. Acute Toxoplasmosis in Squirrel Monkeys (Saimiri Sciureus) in Mexico. *Veterinary Parasitology* 180: 368–71. [CrossRef]

Cermáková, Z., Olga Rysková, and L. Plísková. 2005. Polymerase Chain Reaction for Detection of *Toxoplasma gondii* in Human Biological Samples. *Folia Microbiologica* 50: 341–44. [CrossRef]

Chaklu, Meselu, Zewdu Seyoum Tarekegn, Girma Birhan, and Shimelis Dagnachew. 2020. *Toxoplasma gondii* Infection in Backyard Chickens (Gallus Domesticus): Seroprevalence and Associated Risk Factors in Northwest Ethiopia. *Veterinary Parasitology, Regional Studies and Reports* 21: 100425. [CrossRef]

Chapey, Emmanuelle, Valeria Meroni, François Kieffer, Lina Bollani, René Ecochard, Patricia Garcia, Martine Wallon, and François Peyron. 2015. Use of IgG in Oral Fluid to Monitor Infants with Suspected Congenital Toxoplasmosis. *Clinical and Vaccine Immunology: CVI* 22: 398–403. [CrossRef]

Chávez-Velásquez, Amanda, Gema Alvarez-García, Mercedes Gómez-Bautista, Eva Casas-Astos, Enrique Serrano-Martínez, and Luis Miguel Ortega-Mora. 2005. *Toxoplasma gondii* Infection in Adult Llamas (Lama Glama) and Vicunas (Vicugna vicugna) in the Peruvian Andean Region. *Veterinary Parasitology* 130: 93–97. [CrossRef] [PubMed]

Chen, Daixiong, Jiaqing Tan, Ying Hu, Haoxian Shen, Xiaomin Li, Jian Huo, Jianning Liu, Yichai Zhang, and Yueqing Cai. 2005. Epidemiological Investigation of *Toxoplasma gondii* Infection in Cats and Dogs in Guangzhou. *Journal of Tropical Medicine* 5: 639–41.

Chen, Ju-Chi, Yu-Jen Tsai, and Ying-Ling Wu. 2015. Seroprevalence of *Toxoplasma gondii* Antibodies in Wild Birds in Taiwan. *Research in Veterinary Science* 102: 184–88. [CrossRef] [PubMed]

Chiari, C. D. A., and D. P. Neves. 1984. [Human toxoplasmosis acquired by ingestion of goat's milk]. *Memorias Do Instituto Oswaldo Cruz* 79: 337–40. [CrossRef]

Chinchilla, Misael, Olga Marta Guerrero, A. Castro, and J. Sabah. 1994. Cockroaches as Transport Hosts of the Protozoan *Toxoplasma gondii*. *Revista De Biologia Tropical* 42: 329–31.

Cilievici, Suzana Elena, and Carmen Michaela Creţu. 2007. Strategii de Diagnostic În Toxoplasmoza Umană. *Revista Română de Parazitologie* 17: 104.

Clementino, Maysa Mandetta, Maria Fatima Souza, and Valter Fereira Andrade Neto. 2007. Seroprevalence and *Toxoplasma gondii*-IgG Avidity in Sheep from Lajes, Brazil. *Veterinary Parasitology* 146: 199–203. [CrossRef]

Coelho, A. M., G. Cherubini, A. De Stefani, A. Negrin, Rodrigo Gutierrez-Quintana, Erika Bersan, and Julien Guevar. 2019. Serological Prevalence of Toxoplasmosis and Neosporosis in Dogs Diagnosed with Suspected Meningoencephalitis in the UK. *The Journal of Small Animal Practice* 60: 44–50. [CrossRef] [PubMed]

Coelho, Catarina, Madalena Vieira-Pinto, Ana Sofia Faria, Hélia Vale-Gonçalves, Octávia Veloso, Maria das Neves Paiva-Cardoso, João Rodrigo Mesquita, and Ana Patrícia Lopes. 2014. Serological Evidence of *Toxoplasma gondii* in Hunted Wild Boar from Portugal. *Veterinary Parasitology* 202: 310–12. [CrossRef]

Colville, Joann, and David Berryhill. 2007. *Handbook of Zoonoses: Identification and Prevention*, 1st ed. St. Louis: Mosby.

Cong, Wei, Hany M. Elsheikha, Na Zhou, Peng Peng, Si-Yuan Qin, Qing-Feng Meng, and Ai-Dong Qian. 2018. Prevalence of Antibodies against *Toxoplasma gondii* in Pets and Their Owners in Shandong Province, Eastern China. *BMC Infectious Diseases* 18: 430. [CrossRef]

Conrad, Patricia A., Melissa Ann Miller, Christine Kreuder, Erick R. James, Joruia Mazet, Haydee Dabritz, David A. Jessup, Frances Gulland, and Michael E. Grigg. 2005. Transmission of Toxoplasma: Clues from the Study of Sea Otters as Sentinels of *Toxoplasma gondii* Flow into the Marine Environment. *International Journal for Parasitology* 35: 1155–68. [CrossRef] [PubMed]

Contini, Carlo, Silva Seraceni, Rosario Cultrera, Carlo Incorvaia, Adolfo Sebastiani, and Stephan Picot. 2005. Evaluation of a Real-Time PCR-Based Assay Using the Lightcycler System for Detection of *Toxoplasma gondii* Bradyzoite Genes in Blood Specimens from Patients with Toxoplasmic Retinochoroiditis. *International Journal for Parasitology* 35: 275–83. [CrossRef] [PubMed]

Cook, Alasdair J., R. E. Gilbert, Wilma Buffolano, Jade Zufferey, Eskild Petersen, Pal A. Jenum, W. Foulon, Augusto E. Semprini, and David T. Dunn. 2000. Sources of Toxoplasma Infection in Pregnant Women: European Multicentre Case-Control Study. European Research Network on Congenital Toxoplasmosis. *BMJ (Clinical Research Ed.)* 321: 142–47. [CrossRef] [PubMed]

Correa, Ricardo, Ivonne Cedeño, Cecilia de Escobar, and Isabel Fuentes. 2008. Increased Urban Seroprevalence of *Toxoplasma gondii* Infecting Swine in Panama. *Veterinary Parasitology* 153: 9–11. [CrossRef] [PubMed]

Cosoroabă, Iustin. 2005. *Zoonoze Parazitare.* Timişoara: First/Artpress.

Costa, D. G. C., Maria Fernanda Vianna Marvulo, J. S. A. Silva, S. C. Santana, Fernando Jorge Rodrigues Magalhães, Carlos Diogenes Ferreira de Lima Filho, and Vanessa Oliviera Ribeiro. 2012. Seroprevalence of *Toxoplasma gondii* in Domestic and Wild Animals from the Fernando de Noronha, Brazil. *The Journal of Parasitology* 98: 679–80. [CrossRef] [PubMed]

Costa, Gustavo Henrique Nogueira, Alvimar José da Costa, Welber Daniel Zanetti Lopes, Katia Denise Saraiva Bresciani, Thais Rabelo dos Santos, César Roberto Esper, and Aureo Evangelista Santana. 2011. *Toxoplasma gondii*: Infection Natural Congenital in Cattle and an Experimental Inoculation of Gestating Cows with Oocysts. *Experimental Parasitology* 127: 277–81. [CrossRef]

Cozma, Vasile, O. Şuteu, A. Titilincu, and R.M. Osztián. 2007. Seroprevalenţa Anticorpilor Anti-*Toxoplasma gondii* La Câinii Comunitari Din Cluj-Napoca. *Revista Română de Parazitologie* 17: 23–26.

Cozon, G. J., J. Ferrandiz, H. Nebhi, Martine Wallon, and Francois Peyron. 1998. Estimation of the Avidity of Immunoglobulin G for Routine Diagnosis of Chronic *Toxoplasma gondii* Infection in Pregnant Women. *European Journal of Clinical Microbiology & Infectious Diseases: Official Publication of the European Society of Clinical Microbiology* 17: 32–36. [CrossRef]

Craver, Mary Patricia J., Peggy J. Rooney, and Laura J. Knoll. 2010. Isolation of *Toxoplasma gondii* Development Mutants Identifies a Potential Proteophosphogylcan That Enhances Cyst Wall Formation. *Molecular and Biochemical Parasitology* 169: 120–23. [CrossRef] [PubMed]

Czopowicz, Michał, Jarosław Kaba, Olga Szaluś-Jordanow, Mariusz Nowicki, Lucjan Witkowski, and Tadeusz Frymus. 2011. Seroprevalence of *Toxoplasma gondii* and Neospora Caninum Infections in Goats in Poland. *Veterinary Parasitology* 178: 339–41. [CrossRef] [PubMed]

da Cunha, Graziela Ribeiro, Maysa Pellizzaro, Camila Marinelli Martins, Suzana Maria Rocha, Ana Carolina Yamakawa, Evelyn Cristine da Silva, Andrea Pires Dos Santos, Vivien Midori Morikawa, Hélio Langoni, and Alexander Welker Biondo. 2020. Spatial Serosurvey of Anti-*Toxoplasma gondii* Antibodies in Individuals with Animal Hoarding Disorder and Their Dogs in Southern Brazil. *PLoS ONE* 15: e0233305. [CrossRef]

da Silva, Aristeu Vieira, Sandia Bergamaschi Pezerico, Vanessa Yuri de Lima, Leandro d'Arc Moretti, Julia Plombom Pinheiro, Erica Maeme Tanaka, Márcio Garcia Ribeiro, and Helio Langoni. 2005. Genotyping of *Toxoplasma gondii* Strains Isolated from Dogs with Neurological Signs. *Veterinary Parasitology* 127: 23–27. [CrossRef]

da Silva, Rodrigo Costa, and Helio Langoni. 2016. Risk Factors and Molecular Typing of *Toxoplasma gondii* Isolated from Ostriches (Struthio Camelus) from a Brazilian Slaughterhouse. *Veterinary Parasitology* 225: 73–80. [CrossRef] [PubMed]

da Silva, Rodrigo Costa, Carolina Ballarini Zetun, Sandra de Moraes Gimenes Bosco, Eduardo Bagagli, Patrícia Sammarco Rosa, and Hélio Langoni. 2008. *Toxoplasma gondii* and Leptospira Spp. Infection in Free-Ranging Armadillos. *Veterinary Parasitology* 157: 291–93. [CrossRef] da Silva, Rodrigo Costa, Gustavo Puglia Machado, Tatiane Morosini de Andrade Cruvinel, Ciro Alexandre Cruvinel, and Helio Langoni. 2014. Detection of Antibodies to *Toxoplasma gondii* in Wild Animals in Brazil. *The Journal of Venomous Animals and Toxins including Tropical Diseases* 20: 41. [CrossRef] [PubMed]

da Silva, Rodrigo Costa, Helio Langoni, Chunlei Su, and Aristeu Vieira da Silva. 2011. Genotypic Characterization of *Toxoplasma gondii* in Sheep from Brazilian Slaughterhouses: New Atypical Genotypes and the Clonal Type II Strain Identified. *Veterinary Parasitology* 175: 173–77. [CrossRef]

Daggett, Pierre-Marc, and Thomas A. Nerad. 1992. Cryopreservation of *Toxoplasma gondii*. *Protocols in Protozoology (Society of Protozoologists)* A: 85.1–85.3. Available online: https://protistologists.org/wp-content/uploads/2016/06/Protocols-in-protozoology.pdf (accessed on 10 November 2020).

Dahmani, Ali, Khaled Harhoura, Miriem Aissi, Safia Zenia, B. Hamriouri, N. Guechi, M. Ait Athmane, and R. Kadour. 2018. The Zoonotic Protozoan of Sheep Carcasses in the North of Algeria: A Case of Ovine Toxoplasmosis. *Journal of the Hellenic Veterinary Medical Society* 69: 1004. [CrossRef]

Dahourou, Laibané Dieudonné, Oubri Bassa Gbati, Madi Savadogo, Bernadette Yougbare, Amadou Dicko, Alima Hadjia Banyala Combari, and Alain Richi Kamga-Waladjo. 2019. Prevalence of *Toxoplasma gondii* and Neospora Caninum Infections in Households Sheep "Elevage En Case" in Dakar, Senegal. *Veterinary World* 12: 1028–32. [CrossRef]

Damriyasa, I. M., and Christian Bauer. 2005. Seroprevalence of *Toxoplasma gondii* Infection in Sows in Münsterland, Germany. *DTW. Deutsche Tierarztliche Wochenschrift* 112: 223–24.

Damriyasa, I. Made, Christian Bauer, Renate Edelhofer, Klaus Failing, Peter Lind, Eskild Petersen, Gereon Schares, Astrid Margareta Tenter, R. Volmer, and Horst Zahner. 2004. Cross-Sectional Survey in Pig Breeding Farms in Hesse, Germany: Seroprevalence and Risk Factors of Infections with *Toxoplasma gondii*, Sarcocystis Spp. and Neospora Caninum in Sows. *Veterinary Parasitology* 126: 271–86. [CrossRef] [PubMed]

Dangolla, Ashoka, D. K. Ekanayake, Rajapakse Peramune Vedikkarage Jayanthe Rajapakse, Jitender P. Dubey, and Indira D. Silva. 2006. Seroprevalence of *Toxoplasma gondii* Antibodies in Captive Elephants (Elephaus Maximus Maximus) in Sri Lanka. *Veterinary Parasitology* 137: 172–74. [CrossRef] [PubMed]

Dărăbuş, G., I. Hotea, I. Oprescu, S. Morariu, I. Brudiu, and R. T. Olariu. 2011a. Toxoplasmosis Seroprevalence in Cats and Sheep from Western Romania. *Revue de Medecine Veterinaire.* 162: 316–20.

Dărăbuş, Gheorghe, Mihăiţă Afrenie, Rareş Olariu, Marius Ilie, Adrian Balint, and Ionela Hotea. 2011b. Epidemiological Remarks on *Toxoplasma gondii* Infection in Timişoara Zoo. *Scientia Parasitologica* 12: 33–37.

Dărăbuş, Gheorghe, Ion Oprescu, Sorin Morariu, and Narcisa Mederle. 2006. *Puruzitologie şi Boli Parazitare.* Timişoara: Mirton.

Davies, Peter Robert, William Edward Morrow, John Deen, H. Ray Gamble, and Sharon Patton. 1998. Seroprevalence of *Toxoplasma gondii* and *Trichinella spiralis* in Finishing Swine Raised in Different Production Systems in North Carolina, USA. *Preventive Veterinary Medicine* 36: 67–76. [CrossRef] [PubMed]

Davoust, B., O. Mediannikov, Cedric Roqueplo, C. Perret, Jean Paul Demoncheaux, Masse Sambou, Jacques Guillot, and Radu Blaga. 2015. Enquête de séroprévalence de la toxoplasmose animale au Sénégal. *Bulletin de la Société de Pathologie Exotique* 108: 73–77. [CrossRef] de A Dos Santos, Cláudia B., Angela C. F. B. de Carvalho, Alessandra M. A. Ragozo, Rodrigo M. Soares, Marcos Amaku, Lúcia E. O. Yai, Jitender P. Dubey, and Solange M. Gennari. 2005. First Isolation and Molecular Characterization of *Toxoplasma gondii* from Finishing Pigs from São Paulo State, Brazil. *Veterinary Parasitology* 131: 207–11. [CrossRef]

de Brito, Adriana Falco, Luiz Carlos de Souza, Aristeu Vieira da Silva, and Helio Langoni. 2002. Epidemiological and Serological Aspects in Canine Toxoplasmosis in Animals with Nervous Symptoms. *Memorias Do Instituto Oswaldo Cruz* 97: 31–35. [CrossRef]

De Craeye, Stephane, Niko Speybroeck, Daniel Ajzenberg, Marie Laure Dardé, Frederic Collinet, Paul Tavernier, Steven Van Gucht, Pierre Dorny, and Katelijne Dierick. 2011. *Toxoplasma gondii* and Neospora Caninum in Wildlife: Common Parasites in Belgian Foxes and Cervidae? *Veterinary Parasitology* 178: 64–69. [CrossRef] [PubMed]

De Craeye, Stéphane, Aurelie Francart, Julie Chabauty, Veerle De Vriendt, Steven Van Gucht, Ingrid Leroux, and Erik Jongert. 2008. Prevalence of *Toxoplasma gondii* Infection in Belgian House Cats. *Veterinary Parasitology* 157: 128–32. [CrossRef] [PubMed]

de Oliveira, Gabriela Mota Sena, Juçara Magalhães Simões, Robert Eduard Schaer, Songeli Menezes Freire, Roberto José Meyer Nascimento, Adélia Maria Carvalho de Melo Pinheiro, Silvia Maria Santos Carvalho, Ana Paula Melo Mariano, Rosely Cabral de Carvalho, and Alexandre Dias Munhoz. 2019. Frequency and Factors Associated with *Toxoplasma gondii* Infection in Pregnant Women and Their Pets in Ilhéus, Bahia, Brazil. *Revista Da Sociedade Brasileira De Medicina Tropical* 52: e20190250. [CrossRef] [PubMed]

de Seabra, Nathália Mendonça, Vanessa Figueredo Pereira, Marcos Vinícius Kuwassaki, Julia Cristina Benassi, and Trícia Maria Ferreira de Sousa Oliveira. 2015. *Toxoplasma gondii*, *Neospora caninum* and *Leishmania* Spp. Serology and *Leishmania* Spp. PCR in Dogs from Pirassununga, SP. *Revista Brasileira De Parasitologia Veterinaria = Brazilian Journal of Veterinary Parasitology: Orgao Oficial Do Colegio Brasileiro De Parasitologia Veterinaria* 24: 454–58. [CrossRef]

de Sousa, Susana, D. Ajzenberg, Nuno Canada, Lina Freire, J. M. Correia da Costa, M. L. Dardé, P. Thulliez, and J. P. Dubey. 2006. Biologic and Molecular Characterization of *Toxoplasma gondii* Isolates from Pigs from Portugal. *Veterinary Parasitology* 135: 133–36. [CrossRef]

de Souza, Soraia Figueiredo, Luciana Dos Santos Medeiros, Adriane De Souza Belfort, Andrey Luiz Lopes Cordeiro, Michelle Federle, Antonio Pereira de Souza, and Anderson Barbosa de Moura. 2015. Anticorpos Anti-*Toxoplasma gondii* Em Gatos Domiciliados No Município de Rio Branco, Acre, Brasil. *Semina: Ciências Agrárias* 36: 3757. [CrossRef]

Deem, Sharon L., Jane Merkel, Lora Ballweber, F. Hernan Vargas, Marilyn B. Cruz, and Patricia G. Parker. 2010. Exposure to *Toxoplasma gondii* in Galapagos Penguins (Spheniscus Mendiculus) and Flightless Cormorants (Phalacrocorax Harrisi) in the Galapagos Islands, Ecuador. *Journal of Wildlife Diseases* 46: 1005–11. [CrossRef] [PubMed]

DeFeo, Monica L., Jitender P. Dubey, Thomas N. Mather, and Richard C. Rhodes. 2002. Epidemiologic Investigation of Seroprevalence of Antibodies to *Toxoplasma gondii* in Cats and Rodents. *American Journal of Veterinary Research* 63: 1714–17. [CrossRef] [PubMed]

Deksne, Gunita, and Muza Kirjušina. 2013. Seroprevalence of *Toxoplasma gondii* in Domestic Pigs (Sus Scrofa Domestica) and Wild Boars (Sus Scrofa) in Latvia. *The Journal of Parasitology* 99: 44–47. [CrossRef]

Deksne, Gunita, Bettija Ligere, Aija Šneidere, and Pikka Jokelainen. 2017. Seroprevalence and Factors Associated with *Toxoplasma gondii* Infections in Sheep in Latvia: Latvian Dark Headed Sheep Breed Associated with Higher Seroprevalence. *Vector Borne and Zoonotic Diseases (Larchmont, NY)* 17: 478–82. [CrossRef] [PubMed]

Demissie, T., and G. Tilahun. 2004. Study on Toxoplasmosis in Sheep and Goats in Debre Birhan and the Surrounding Areas in Ethiopia. *Veterinary Bulletin* 74: 646.

de Moura, Anderson B., Silvia C. Osaki, Dauton L. Zulpo, and Elizabete R. M. Marana. 2007. [Occurrence of anti-Toxoplasma gondii antibodies in swine and ovine slaughtered at municipality of Guarapuava in the State of Paraná, Brazil]. *Revista Brasileira De Parasitologia Veterinaria = Brazilian Journal of Veterinary Parasitology: Orgao Oficial Do Colegio Brasileiro De Parasitologia Veterinaria* 16: 54–56.

Deng, Genny, L. Ou, X. Tang, and W. Wang. 2011. Serological Survey of Toxoplasmosis in Dogs and Cats in Changsha City. *Chinese Journal of Veterinary Medicine* 50: 74–75.

Desmonts, Georges, and Jacques Couvreur. 1974. Congenital Toxoplasmosis. A Prospective Study of 378 Pregnancies. *The New England Journal of Medicine* 290: 1110–16. [CrossRef] [PubMed]

DeSouza, Silvio L.P., Solange Gennari, Lucia E.O. Yai, Sandra R.N. D'Auria, Sonia M.S. Cardoso, Jose S.G. Junior, and Jitender P. Dubey. 2003. Occurrence of *Toxoplasma gondii* Antibodies in Sera from Dogs of the Urban and Rural Areas from Brazil. *Revista Brasileira de Parasitologia Veterinaria* 12: 1–3.

Devleesschauwer, Brecht, Mathieu Pruvot, Durga Datt Joshi, Stéphane De Craeye, Malgorzata Jennes, Anita Ale, and Alma Welinski. 2013. Seroprevalence of Zoonotic Parasites in Pigs Slaughtered in the Kathmandu Valley of Nepal. *Vector Borne and Zoonotic Diseases (Larchmont, N.Y.)* 13: 872–76. [CrossRef] [PubMed]

Diakoua, Anastasia, Elias Papadopoulos, Nikolaos Panousis, Charilaos Karatzias, and Nektarios Giadinis. 2013. *Toxoplasma gondii* and Neospora Caninum Seroprevalence in Dairy Sheep and Goats Mixed Stock Farming. *Veterinary Parasitology* 198: 387–90. [CrossRef] [PubMed]

Djokic, Vitomir, Ivana Klun, Vincenzo Musella, Laura Rinaldi, Giuseppe Cringoli, Smaragda Sotiraki, and Olgica Djurkovic-Djakovic. 2014. Spatial Epidemiology of *Toxoplasma gondii* Infection in Goats in Serbia. *Geospatial Health* 8: 479–88. [CrossRef]

Djurković-Djaković, Olgica, Aleksandra Nikolić, Branko Bobić, Ivana Klun, and Anastasija Aleksić. 2005. Stage Conversion of *Toxoplasma gondii* RH Parasites in Mice by Treatment with Atovaquone and Pyrrolidine Dithiocarbamate. *Microbes and Infection* 7: 49–54. [CrossRef]

dos Santos, Thaís Rabelo, Cáris Maroni Nunes, Maria Cecília Rui Luvizotto, Anderson Barbosa de Moura, Welber Daniel Zanetti Lopes, Alvimar José da Costa, and Katia Denise Saraiva Bresciani. 2010. Detection of *Toxoplasma gondii* Oocysts in Environmental Samples from Public Schools. *Veterinary Parasitology* 171: 53–57. [CrossRef] [PubMed]

Dubey, Jitender P. 2000. The Scientific Basis for Prevention of *Toxoplasma gondii* Infection: Studies on Tissue Cyst Survival, Risk Factors and Hygiene Measures. In *Congenital Toxoplasmosis*. Edited by Pierre Ambroise-Thomas and Peter Eskild Petersen. Paris: Springer, pp. 271–75. [CrossRef]

Dubey, Jitender P. 2002. A Review of Toxoplasmosis in Wild Birds. *Veterinary Parasitology* 106: 121–53. [CrossRef] [PubMed]

Dubey, Jitender P. 2004. Toxoplasmosis—A Waterborne Zoonosis. *Veterinary Parasitology* 126: 57–72. [CrossRef]

Dubey, Jitender P. 2005. Unexpected Oocyst Shedding by Cats Fed *Toxoplasma gondii* Tachyzoites: In Vivo Stage Conversion and Strain Variation. *Veterinary Parasitology* 133: 289–98. [CrossRef]

Dubey, Jitender P. 2006. Comparative Infectivity of Oocysts and Bradyzoites of *Toxoplasma gondii* for Intermediate (Mice) and Definitive (Cats) Hosts. *Veterinary Parasitology* 140: 69–75. [CrossRef] [PubMed]

Dubey, Jitender P. 2009a. History of the Discovery of the Life Cycle of *Toxoplasma gondii*. *International Journal for Parasitology* 39: 877–82. [CrossRef] [PubMed]

Dubey, Jitender P. 2009b. Toxoplasmosis in Sheep—The Last 20 Years. *Veterinary Parasitology* 163: 1–14. [CrossRef] [PubMed]

Dubey, Jitender P. 2009c. Toxoplasmosis in Pigs--the Last 20 Years. *Veterinary Parasitology* 164: 89–103. [CrossRef]

Dubey, Jitender P. 2010. *Toxoplasmosis of Animals and Humans.* Boca Raton: CRC Press.

Dubey, Jitender P., and Colin P. Beattie. 1990. Toxoplasmosis of Animals and Man. By Jitender P. Dubey and Colin P. Beattie. 220 Pages. ISBN 0 8493 4618 5. CRC Press, Boca Raton, 1988. £108.00. *Parasitology* 100: 500–1. [CrossRef]

Dubey, Jitender P., and An Pas. 2008. *Toxoplasma gondii* Infection in Blanford's Fox (Vulpes Cana). *Veterinary Parasitology* 153: 147–51. [CrossRef]

Dubey, Jitender P., and Jeffrey L. Jones. 2008. *Toxoplasma gondii* Infection in Humans and Animals in the United States. *International Journal for Parasitology* 38: 1257–78. [CrossRef] [PubMed]

Dubey, Jitender P., and M. R. Lappin. 1998. Toxoplasmosis and Neosporosis. In *Infectious Diseases of the Dog and Cat*, 2nd ed. Edited by C. E. Greene. Philadelphia: W.B. Saunders Company.

Dubey, Jitender P., Barbara Lewis, K. Beam, and Bruce Abbitt. 2002a. Transplacental Toxoplasmosis in a Reindeer (Rangifer Tarandus) Fetus. *Veterinary Parasitology* 110: 131–35. [CrossRef] [PubMed]

Dubey, Jitender P., Claudia Ehlers Kerber, and David E. Granstrom. 1999a. Serologic Prevalence of Sarcocystis Neurona, *Toxoplasma gondii*, and Neospora Caninum in Horses in Brazil. *Journal of the American Veterinary Medical Association* 215: 970–72. [PubMed]

Dubey, Jitender P., C. Rajendran, D. G. C. Costa, L. R. Ferreira, Oliver Chun Hung Kwok, D. Qu, and Chunlei Su. 2010a. New *Toxoplasma gondii* Genotypes Isolated from Free-Range Chickens from the Fernando de Noronha, Brazil: Unexpected Findings. *The Journal of Parasitology* 96: 709–12. [CrossRef]

Dubey, Jitender P., Chellaiah Rajendran, L. R. Ferreira, Oliver Chun Hung Kwok, D. Sinnett, D. Majumdar, and Chunlei Su. 2010b. A New Atypical Highly Mouse Virulent *Toxoplasma gondii* Genotype Isolated from a Wild Black Bear in Alaska. *The Journal of Parasitology* 96: 713–16. [CrossRef] [PubMed]

Dubey, Jitender P., Chunlei Su, Jesus Alfredo Cortés, Natarajan Sundar, Jorge Enrique Gomez-Marin, Luis J. Polo, L. Zambrano, L.E. Mora, Fabiana Lora-Suarez, J. Jimenez, and et al. 2006a. Prevalence of *Toxoplasma gondii* in Cats from Colombia, South America and Genetic Characterization of *T. gondii* Isolates. *Veterinary Parasitology* 141: 42–47. [CrossRef] [PubMed]

Dubey, Jitender P., Chunlei Su, J. Oliveira, Juan Alberto Morales, Raquel V. Bolaños, Natarajan Sundar, Oliver Chun Hung Kwok, and S. K. Shen. 2006b. Biologic and Genetic Characteristics of *Toxoplasma gondii* Isolates in Free-Range Chickens from Costa Rica, Central America. *Veterinary Parasitology* 139: 29–36. [CrossRef]

Dubey, Jitender P., D. E. Hill, Jeffrey L. Jones, A. W. Hightower, E. Kirkland, J. M. Roberts, Paula L. Marcet, T. Lehmann, Manoella Campostrini Baretto Vianna, Katarzyna Miska, and et al. 2005a. Prevalence of Viable *Toxoplasma gondii* in Beef, Chicken, and Pork from Retail Meat Stores in the United States: Risk Assessment to Consumers. *The Journal of Parasitology* 91: 1082–93. [CrossRef] [PubMed]

Dubey, Jitender P., Douglas H. Graham, C. R. Blackston, T. Lehmann, Solange Maria Gennari, Alessandra Mara Alves Ragozo, and Sandra M. Nishi. 2002b. Biological and Genetic Characterisation of *Toxoplasma gondii* Isolates from Chickens (Gallus Domesticus) from São Paulo, Brazil: Unexpected Findings. *International Journal for Parasitology* 32: 99–105. [CrossRef] [PubMed]

Dubey, Jitender P., Douglas H. Graham, R. W. De Young, E. Dahl, M. L. Eberhard, Eva K. Nace, and K. Won. 2004a. Molecular and Biologic Characteristics of *Toxoplasma gondii* Isolates from Wildlife in the United States. *The Journal of Parasitology* 90: 67–71. [CrossRef]

Dubey, Jitender P., Dale M. Webb, Natarajan Sundar, Gopal Viswanathan Velmurugan, Luciana A. Bandini, Oliver Chun Hung Kwok, and Chunlei Su. 2007a. Endemic Avian Toxoplasmosis on a Farm in Illinois: Clinical Disease, Diagnosis, Biologic and Genetic Characteristics of *Toxoplasma gondii* Isolates from Chickens (Gallus Domesticus), and a Goose (Anser Anser). *Veterinary Parasitology* 148: 207–12. [CrossRef]

Dubey, Jitender P., David S. Lindsay, William J. Saville, Stephen M. Reed, David E. Granstrom, and C. A. Speer. 2001. A Review of Sarcocystis Neurona and Equine Protozoal Myeloencephalitis (EPM). *Veterinary Parasitology* 95: 89–131. [CrossRef]

Dubey, Jitender P., Edward A. Rollor, K. Smith, Oliver Chun Kwok, and Philippe Thulliez. 1997. Low Seroprevalence of *Toxoplasma gondii* in Feral Pigs from a Remote Island Lacking Cats. *The Journal of Parasitology* 83: 839–41. [CrossRef]

Dubey, Jitender P., E. Clay Hodgin, and A. N. Hamir. 2006c. Acute Fatal Toxoplasmosis in Squirrels (Sciurus Carolensis) with Bradyzoites in Visceral Tissues. *The Journal of Parasitology* 92: 658–59. [CrossRef]

Dubey, Jitender P., E. S. Morales, and T. Lehmann. 2004b. Isolation and Genotyping of *Toxoplasma gondii* from Free-Ranging Chickens from Mexico. *The Journal of Parasitology* 90: 411–13. [CrossRef]

Dubey, Jitender P., Gopal Viswanathan Velmurugan, A. Chockalingam, Hilda Fatima Jesus Pena, L. Nunes de Oliveira, C. A. Leifer, Solange Maria Gennari, Lilian Maria Garcia Bahia Oliveira, and Chunlei Su. 2008a. Genetic Diversity of *Toxoplasma gondii* Isolates from Chickens from Brazil. *Veterinary Parasitology* 157: 299–305. [CrossRef]

Dubey, Jitender P., Gopal Viswanthan Velmurugan, Chellaiah Rajendran, Michael J. Yabsley, Nancy J. Thomas, Kimberlee B. Beckmen, D. Sinnett, David Ruid, J. Hart, Patricia Fair, and et al. 2011. Genetic Characterisation of *Toxoplasma gondii* in Wildlife from North America Revealed Widespread and High Prevalence of the Fourth Clonal Type. *International Journal for Parasitology* 41: 1139–47. [CrossRef]

Dubey, Jitender P., Gopal Viswanthan Velmurugan, V. Ulrich, James Gill, Michelle Carstensen, Natarajan Sundar, Oliver Chun Hung Kwok, Philippe Thulliez, Debashree Majumdar, and Chunlei Su. 2008b. Transplacental Toxoplasmosis in Naturally-Infected White-Tailed Deer: Isolation and Genetic Characterisation of *Toxoplasma gondii* from Foetuses of Different Gestational Ages. *International Journal for Parasitology* 38: 1057–63. [CrossRef] [PubMed]

Dubey, Jitender P., H. Ray Gamble, D. Hill, Chirukandoth Sreekumar, Stephane Romand, and P. Thuilliez. 2002c. High Prevalence of Viable *Toxoplasma gondii* Infection in Market Weight Pigs from a Farm in Massachusetts. *The Journal of Parasitology* 88: 1234–38. [CrossRef] [PubMed]

Dubey, Jitender P., Ionela Hotea, T. R. Olariu, J. L. Jones, and Gheorghe Dărăbuş. 2014a. Epidemiological Review of Toxoplasmosis in Humans and Animals in Romania. *Parasitology* 141: 311–25. [CrossRef] [PubMed]

Dubey, Jitender P., Italmar Teodorico Navarro, C. Sreekumar, E. Dahl, Roberta Lemos Freire, H. H. Kawabata, and M. C. B. Vianna. 2004c. *Toxoplasma gondii* Infections in Cats from Paraná, Brazil: Seroprevalence, Tissue Distribution, and Biologic and Genetic Characterization of Isolates. *The Journal of Parasitology* 90: 721–26. [CrossRef] [PubMed]

Dubey, Jitender P., Italmar Teodorico Navarro, D. H. Graham, E. Dahl, Roberta Lemos Freire, L. B. Prudencio, Chirukandoth Sreekumar, M. C. Vianna, and Tovi Lehmann. 2003b. Characterization of *Toxoplasma gondii* Isolates from Free Range Chickens from Paraná, Brazil. *Veterinary Parasitology* 117: 229–34. [CrossRef]

Dubey, Jitender P., Jesus Alfredo Cortés-Vecino, Jimmy J. Vargas-Duarte, Natarajan Sundar, Gopal Viswanathan Velmurugan, Luciana M. Bandini, Luis J. Polo, L. Zambrano, L.E. Mora, Oliver Chun Hung Kwok, and et al. 2007b. Prevalence of *Toxoplasma gondii* in Dogs from Colombia, South America and Genetic Characterization of *T. gondii* Isolates. *Veterinary Parasitology* 145: 45–50. [CrossRef]

Dubey, Jitender P., Justin Brown, Shiv Kumar Verma, Camila Koutsodontis Cerqueira-Cézar, Jeremy Banfield, Oliver Chun Hung Kwok, Yuqing Ying, Fernando Henrique Antunes Murata, Abani Kumar Pradhan, and Chunlei Su. 2017. Isolation of Viable *Toxoplasma gondii*, Molecular Characterization, and Seroprevalence in Elk (Cervus Canadensis) in Pennsylvania, USA. *Veterinary Parasitology* 243: 1–5. [CrossRef]

Dubey, Jitender P., J. C. Leighty, V. C. Beal, W. R. Anderson, C. D. Andrews, and Philippe Thulliez. 1991. National Seroprevalence of *Toxoplasma gondii* in Pigs. *The Journal of Parasitology* 77: 517–21. [CrossRef]

Dubey, Jitender P., Jennifer L. Chapman, Benjamin M. Rosenthal, Mark Mense, and Ronald L. Schueler. 2006d. Clinical Sarcocystis Neurona, Sarcocystis Canis, *Toxoplasma gondii*, and Neospora Caninum Infections in Dogs. *Veterinary Parasitology* 137: 36–49. [CrossRef]

Dubey, Jitender P., Kristin Mansfield, B. Hall, Oliver Chun Hung Kwok, and Philippe Thulliez. 2008c. Seroprevalence of Neospora Caninum and *Toxoplasma gondii* in Black-Tailed Deer (Odocoileus Hemionus Columbianus) and Mule Deer (Odocoileus Hemionus Hemionus). *Veterinary Parasitology* 156: 310–13. [CrossRef]

Dubey, Jitender P., Lam Thi Thu Huong, Natarajan Sundar, and Chunlei Su. 2007c. Genetic Characterization of *Toxoplasma gondii* Isolates in Dogs from Vietnam Suggests Their South American Origin. *Veterinary Parasitology* 146: 347–51. [CrossRef] [PubMed]

Dubey, Jitender P., Manoella Campostrini Barreto Vianna, Susana Sousa, Nuna Canada, Silvia Meireles, Jose Manuel Correia da Costa, Paula L. Marcet, Tovi Lehmann, Marie Laure Dardé, and Philippe Thulliez. 2006e. Characterization of *Toxoplasma gondii* Isolates in Free-Range Chickens from Portugal. *The Journal of Parasitology* 92: 184–86. [CrossRef] [PubMed]

Dubey, Jitender P., Mark C. Jenkins, Oliver Chun Hung Kwok, Rebecca L. Zink, Michelle L. Michalski, V. Ulrich, James Gill, Michelle Carstensen, and Philippe Thulliez. 2009. Seroprevalence of Neospora Caninum and *Toxoplasma gondii* Antibodies in White-Tailed Deer (Odocoileus Virginianus) from Iowa and Minnesota Using Four Serologic Tests. *Veterinary Parasitology* 161: 330–34. [CrossRef] [PubMed]

Dubey, Jitender P., Maria Cecilia Venturini, Lucila Venturini, J. McKinney, and Marcelo Pecoraro. 1999b. Prevalence of Antibodies to Sarcocystis Neurona, *Toxoplasma gondii* and Neospora Caninum in Horses from Argentina. *Veterinary Parasitology* 86: 59–62. [CrossRef]

Dubey, Jitender P., Michael Z. Levy, Chirukandoth Sreekumar, Oliver Chun Hung Kwok, S. K. Shen, E. Dahl, Philippe Thulliez, and Tovi Lehmann. 2004d. Tissue Distribution and Molecular Characterization of Chicken Isolates of *Toxoplasma gondii* from Peru. *The Journal of Parasitology* 90: 1015–18. [CrossRef]

Dubey, Jitender P., Nandu Sundar, Cherrie A. Nolden, M. D. Samuel, Gopal Viswanthan Velmurugan, Luciana A. Bandini, Oliver Chun Hung Kwok, B. Bodenstein, and Chunlei Su. 2007d. Characterization of *Toxoplasma gondii* from Raccoons (Procyon Lotor), Coyotes (Canis Latrans), and Striped Skunks (Mephitis Mephitis) in Wisconsin Identified Several Atypical Genotypes. *The Journal of Parasitology* 93: 1524–27. [CrossRef]

Dubey, Jitender P., Natarajan Sundar, Dolores Hill, Gopal Viswanathan Velmurugan, Luciana A. Bandini, Oliver Chun Hung Kwok, Debashree Majumdar, and Chunlei Su. 2008d. High Prevalence and Abundant Atypical Genotypes of *Toxoplasma gondii* Isolated from Lambs Destined for Human Consumption in the USA. *International Journal for Parasitology* 38: 999–1006. [CrossRef]

Dubey, Jitender P., Natarajan Sundar, N. Pineda, Niels Chr Kyvsgaard, Luz A. Luna, E. Rimbaud, J. B. Oliveira, Oliver Chun Hung Kwok, Y. Qi, and Chunlei Su. 2006f. Biologic and Genetic Characteristics of *Toxoplasma gondii* Isolates in Free-Range Chickens from Nicaragua, Central America. *Veterinary Parasitology* 142: 47–53. [CrossRef]

Dubey, Jitender P., Natarajan Sundar, Solange Maria Gennari, Antonio H. H. Minervino, Nara Amelia da Rosa Farias, Jeronimo Lopes Ruas, Tania Regina Bettin dos Santos, Guacyara Tenorio Cavalcante, Oliver Chun Hung Kwok, and Chunlei Su. 2007e. Biologic and Genetic Comparison of *Toxoplasma gondii* Isolates in Free-Range Chickens from the Northern Pará State and the Southern State Rio Grande Do Sul, Brazil Revealed Highly Diverse and Distinct Parasite Populations. *Veterinary Parasitology* 143: 182–88. [CrossRef]

Dubey, Jitender P., Pamela G. Parnell, Chirukandoth Sreekumar, Manoella Campostrini Barreto Vianna, Randy W. De Young, E. Dahl, and Tovi Lehmann. 2004e. Biologic and Molecular Characteristics of *Toxoplasma gondii* Isolates from Striped Skunk (Mephitis Mephitis), Canada Goose (Branta Canadensis), Black-Winged Lory (Eos Cyanogenia), and Cats (Felis Catus). *The Journal of Parasitology* 90: 1171–74. [CrossRef]

Dubey, Jitender P., Philippe Thulliez, and E. C. Powell. 1995a. *Toxoplasma gondii* in Iowa Sows: Comparison of Antibody Titers to Isolation of *T. gondii* by Bioassays in Mice and Cats. *The Journal of Parasitology* 81: 48–53. [CrossRef] [PubMed]

Dubey, Jitender P., Philippe Thulliez, Stephane Romand, Oliver Chun Hung Kwok, S. K. Shen, and H. Ray Gamble. 1999c. Serologic Prevalence of *Toxoplasma gondii* in Horses Slaughtered for Food in North America. *Veterinary Parasitology* 86: 235–38. [CrossRef] [PubMed]

Dubey, Jitender P., Renate Edelhofer, Paula Marcet, Manoella Campostrini Barreto Vianna, Oliver Chun Hung Kwok, and Tovi Lehmann. 2005b. Genetic and Biologic Characteristics of *Toxoplasma gondii* Infections in Free-Range Chickens from Austria. *Veterinary Parasitology* 133: 299–306. [CrossRef]

Dubey, Jitender P., R. M. Weigel, A. M. Siegel, Philippe Thulliez, Uriel D. Kitron, Mark A. Mitchell, Alessandro Mannelli, Nohra E. Mateus-Pinilla, S. K. Shen, and Oliver Chun Kwok. 1995b. Sources and Reservoirs of *Toxoplasma gondii* Infection on 47 Swine Farms in Illinois. *The Journal of Parasitology* 81: 723–29. [CrossRef] [PubMed]

Dubey, Jitender P., Rajapakse Peramune Vedikkarage Jayanthe Rajapakse, R. R. M. K. Kavindra Wijesundera, Natarajan Sundar, Gopal Viswanathan Velmurugan, Oliver Chun Hung Kwok, and Chunlei Su. 2007f. Prevalence of *Toxoplasma gondii* in Dogs from Sri Lanka and Genetic Characterization of the Parasite Isolates. *Veterinary Parasitology* 146: 341–46. [CrossRef]

Dubey, Jitender P., Randall Zarnke, Nancy J. Thomas, S. K. Wong, William Van Bonn, Michael Briggs, Jay W. Davis, Ruth Ewing, Oliver Chun Hung Kwok, Stephane Romand, and et al. 2003c. *Toxoplasma gondii*, Neospora Caninum, Sarcocystis Neurona, and Sarcocystis Canis-like Infections in Marine Mammals. *Veterinary Parasitology* 116: 275–96. [CrossRef]

Dubey, Jitender P., Sally L. Ness, Oliver Chun Hung Kwok, Shanti Choudhary, Linda D. Mittel, and Thomas J. Divers. 2014b. Seropositivity of *Toxoplasma gondii* in Domestic Donkeys (Equus Asinus) and Isolation of *T. gondii* from Farm Cats. *Veterinary Parasitology* 199: 18–23. [CrossRef]

Dubey, Jitender P., S. M. Mitchell, J. K. Morrow, J. C. Rhyan, L. M. Stewart, D. E. Granstrom, Stephane Romand, Philippe Thulliez, W. J. Saville, and David S. Lindsay. 2003a. Prevalence of Antibodies to Neospora Caninum, Sarcocystis Neurona, and *Toxoplasma gondii* in Wild Horses from Central Wyoming. *The Journal of Parasitology* 89: 716–20. [CrossRef]

Dubey, Jitender P., Solange M. Gennari, Marcelo B. Labruna, Luis M. A. Camargo, Manoella Campostrini Barreto Vianna, Paula L. Marcet, and Tovi Lehmann. 2006g. Characterization of *Toxoplasma gondii* Isolates in Free-Range Chickens from Amazon, Brazil. *The Journal of Parasitology* 92: 36–40. [CrossRef]

Dubey, Jitender P., W. Ben Scandrett, Oliver Chun Kwok, and Alvin A. Gajadhar. 2000. Prevalence of Antibodies to *Toxoplasma gondii* in Ostriches (Struthio Camelus). *The Journal of Parasitology* 86: 623–24. [CrossRef]

Dubey, Jitender P., W. J. A. Saville, J. F. Stanek, and S. M. Reed. 2002d. Prevalence of *Toxoplasma gondii* Antibodies in Domestic Cats from Rural Ohio. *The Journal of Parasitology* 88: 802–3. [CrossRef] [PubMed]

Dubey, Jitender P., Xingquan Q. Zhu, Natarajan Sundar, Han Zhang, Oliver Chun Hung Kwok, and Chunlei Su. 2007g. Genetic and Biologic Characterization of *Toxoplasma gondii* Isolates of Cats from China. *Veterinary Parasitology* 145: 352–56. [CrossRef] [PubMed]

Dubey, Jitender P., Muhammad Iqbal Bhaiyat, Claude de Allie, Calum N. L. Macpherson, Ravindra Nath Sharma, Chirukandoth Sreekumar, Manoella Campostrini Barreto Vianna, S. K. Shen, O. C. H. Kwok, K. B. Miska, and et al. 2005c. Isolation, Tissue Distribution, and Molecular Characterization of *Toxoplasma gondii* from Chickens in Grenada, West Indies. *The Journal of Parasitology* 91: 557–60. [CrossRef] [PubMed]

Dubey, Jitender P., Fernando Henrique Antunes Murata, Camila K. Cerqueira-Cézar, and Oliver Chun Hung Kwok. 2021. Recent Epidemiologic and Clinical *Toxoplasma gondii* Infections in Wild Canids and Other Carnivores: 2009–2020. *Veterinary Parasitology* 290: 109337. [CrossRef] [PubMed]

Dubremetz, Jean François, and David J. P. Ferguson. 2009. The Role Played by Electron Microscopy in Advancing Our Understanding of *Toxoplasma gondii* and Other Apicomplexans. *International Journal for Parasitology* 39: 883–93. [CrossRef]

Dumètre, Aurélien, and Marie Laure Dardé. 2003. How to Detect *Toxoplasma gondii* Oocysts in Environmental Samples? *FEMS Microbiology Reviews* 27: 651–61. [CrossRef]

Dumètre, Aurélien, and Marie-Laure Dardé. 2005. Immunomagnetic Separation of *Toxoplasma gondii* Oocysts Using a Monoclonal Antibody Directed against the Oocyst Wall. *Journal of Microbiological Methods* 61: 209–17. [CrossRef]

Dumètre, Aurélien, and Marie-Laure Dardé. 2007. Detection of *Toxoplasma gondii* in Water by an Immunomagnetic Separation Method Targeting the Sporocysts. *Parasitology Research* 101: 989–96. [CrossRef]

Dumètre, Aurélien, Caroline Le Bras, Maxime Baffet, Pascale Meneceur, Jitender P. Dubey, Francis Derouin, Jean-Pierre Duguet, Michel Joyeux, and Laurent Moulin. 2008. Effects of Ozone and Ultraviolet Radiation Treatments on the Infectivity of *Toxoplasma gondii* Oocysts. *Veterinary Parasitology* 153: 209–13. [CrossRef]

Dumètre, Aurélien, Daniel Ajzenberg, Luc Rozette, Aurélien Mercier, and Marie-Laure Dardé. 2006. *Toxoplasma gondii* Infection in Sheep from Haute-Vienne, France: Seroprevalence and Isolate Genotyping by Microsatellite Analysis. *Veterinary Parasitology* 142: 376–79. [CrossRef]

Dunn, David, Martine Wallon, Francois Peyron, Eskild Petersen, Catherine Peckham, and Ruth Gilbert. 1999. Mother-to-Child Transmission of Toxoplasmosis: Risk Estimates for Clinical Counselling. *Lancet (London, England)* 353: 1829–33. [CrossRef]

Durecko, R., D. Máté, I. Šimanská, and J. Nagy. 2004. Epizootology and Epidemiology of Toxoplasmosis. *Inform. Spracod.* 13: 10–14.

Edelhofer, Renate. 1994. Prevalence of Antibodies against *Toxoplasma gondii* in Pigs in Austria--an Evaluation of Data from 1982 and 1992. *Parasitology Research* 80: 642–44. [CrossRef]

El Behairy, Ahmed M., Shanti Choudhary, Leandra Ribeiro Ferreira, Oliver Chun Hung Kwok, Mosaad Hilali, Chunlei Su, and Jitender P. Dubey. 2013. Genetic Characterization of Viable *Toxoplasma gondii* Isolates from Stray Dogs from Giza, Egypt. *Veterinary Parasitology* 193: 25–29. [CrossRef]

Elfadaly, Hassan A., Nawal A. Hassanain, Raafat M. Shaapan, Mohey A. Hassanain, Ashraf M. Barakat, and Khaled A. Abdelrahman. 2017. Molecular Detection and Genotyping of *Toxoplasma gondii* from Egyptian Isolates. *Asian Journal of Epidemiology* 10: 37–44. [CrossRef]

El-Ghaysh, A. 1998. Seroprevalence of *Toxoplasma gondii* in Egyptian Donkeys Using ELISA. *Veterinary Parasitology* 80: 71–73. [CrossRef]

El-Moukdad, A. R. 2002. [Serological studies on prevalence of Toxoplasma gondii in Awassi sheep in Syria]. *Berliner Und Munchener Tierarztliche Wochenschrift* 115: 186–88.

Enriquez, Gustavo Fabian, Natalia Paula Macchiaverna, Hernan Dario Argibay, Ludmila López Arias, Marisa Farber, Ricado Esteban Gürtler, Marta Victoria Cardinal, and Graciela Garbossa. 2019. Polyparasitism and Zoonotic Parasites in Dogs from a Rural Area of the Argentine Chaco. *Veterinary Parasitology, Regional Studies and Reports* 16: 100287. [CrossRef]

Epiphanio, Sabrina, Idercio Luiz Sinhorini, and Jose Luiz Catão-Dias. 2003. Pathology of Toxoplasmosis in Captive New World Primates. *Journal of Comparative Pathology* 129: 196–204. [CrossRef]

Erkılıç, Ekin Emre, Neriman Mor, Cahit Babur, Ali Haydar Kirmizigul, and Yunus Emre Beyhan. 2016. The Seroprevalence of *Toxoplasma gondii* in Cats from the Kars Region, Turkey. *Israel Journal of Veterinary Medicine* 71: 31–35.

Erscoiu, Simona, E. Ungureanu, D. Alecu, C. Ionescu, and L. Mihaila. 2007. Aspecte Clinice Şi Terapeutice Ale Toxoplasmozei Cerebrale La Adulţi Cu Infecţie HIV. *Revista Română de Parazitologie* 17: 108–9.

Esmerini, Patrícia O., Solange M. Gennari, and Hilda F. J. Pena. 2010. Analysis of Marine Bivalve Shellfish from the Fish Market in Santos City, São Paulo State, Brazil, for *Toxoplasma gondii*. *Veterinary Parasitology* 170: 8–13. [CrossRef]

Esteves, Francisco, Daniela Aguiar, Joana Rosado, Maria Luísa Costa, Bruno de Sousa, Francisco Antunes, and Olga Matos. 2014. *Toxoplasma gondii* Prevalence in Cats from Lisbon and in Pigs from Centre and South of Portugal. *Veterinary Parasitology* 200: 8–12. [CrossRef]

Esubalew, Sisay, Zewdu Seyoum Zed Tarekegn, Wudu Temesgen Jemberu, Shimelis Dagnachew Nigatu, Mulugeta Kussa, Abrham Ayele Tsegaye, Girma Birhan Asteraye, Basazinew Bogale, and Mersha Chanie Kebede. 2020. Seroepidemiology of *Toxoplasma gondii* in Small Ruminants in Northwest Ethiopia. *Veterinary Parasitology, Regional Studies and Reports* 22: 100456. [CrossRef]

Eymann, Jutta, Catherine A. Herbert, Desmond W. Cooper, and J. P. Dubey. 2006. Serologic Survey for *Toxoplasma gondii* and Neospora Caninum in the Common Brushtail Possum (Trichosurus Vulpecula) from Urban Sydney, Australia. *The Journal of Parasitology* 92: 267–72. [CrossRef]

Fábrega, Lorena, Carlos M. Restrepo, Alicia Torres, Diorene Smith, Patricia Chan, Dimas Pérez, Alberto Cumbrera, and Zuleima Caballero E. 2020. Frequency of *Toxoplasma gondii* and Risk Factors Associated with the Infection in Stray Dogs and Cats of Panama. *Microorganisms* 8: 927. [CrossRef]

Fahnehjelm, Kristina Teär, Gunilla Malm, Jan Ygge, Mona-Lisa Engman, Eva Maly, and Birgitta Evengård. 2000. Ophthalmological Findings in Children with Congenital Toxoplasmosis. Report from a Swedish Prospective Screening Study of Congenital Toxoplasmosis with Two Years of Follow-Up. *Acta Ophthalmologica Scandinavica* 78: 569–75. [CrossRef]

Fan, Chia-Kwung, Chien-Ching Hung, Kua-Eyre Su, Hung-Yi Chiou, Vilfrido Gil, Maria da Conceicao dos Reis Ferreira, and Lian-Fen Tseng. 2007. Seroprevalence of *Toxoplasma gondii* Infection among Inhabitants in the Democratic Republic of Sao Tome and Principe. *Transactions of the Royal Society of Tropical Medicine and Hygiene* 101: 1157–58. [CrossRef]

Fan, Chia-Kwung, Kua-Eyre Su, and Yu-Jen Tsai. 2004. Serological Survey of *Toxoplasma gondii* Infection among Slaughtered Pigs in Northwestern Taiwan. *The Journal of Parasitology* 90: 653–54. [CrossRef]

Faria, Eduardo B., Solange M. Gennari, Hilda F. J. Pena, Ana Célia R. Athayde, Maria Luana C. R. Silva, and Sérgio S. Azevedo. 2007. Prevalence of Anti-*Toxoplasma gondii* and Anti-Neospora Caninum Antibodies in Goats Slaughtered in the Public Slaughterhouse of Patos City, Paraíba State, Northeast Region of Brazil. *Veterinary Parasitology* 149: 126–29. [CrossRef]

Fehlhaber, Karsten, Petra Hintersdorf, and Günter Kruger. 2003. Pravalenz von *Toxoplasma gondii*. Untersuchungen Bei Schlachtschtschweinen Aus Verschiedenen Haltungsformen Und in Handelsublichen Hackfleischproben.' Fleischwirtschaft 83: 97–99.

Feitosa, Thais Ferreira, Vinícius Longo Ribeiro Vilela, João Leite de Almeida-Neto, Antonielson Dos Santos, Dayana Firmino de Morais, Bruna Farias Alves, Fabiana Nakashima, Solange Maria Gennari, Ana Célia Rodrigues Athayde, and Hilda Fátima Jesus Pena. 2017. High Genetic Diversity in *Toxoplasma gondii* Isolates from Pigs at Slaughterhouses in Paraíba State, Northeastern Brazil: Circulation of New Genotypes and Brazilian Clonal Lineages. *Veterinary Parasitology* 244: 76–80. [CrossRef]

Felin, Elina, E. Jukola, Saara Raulo, and Maria Fredriksson-Ahomaa. 2015. Meat Juice Serology and Improved Food Chain Information as Control Tools for Pork-Related Public Health Hazards. *Zoonoses and Public Health* 62: 456–64. [CrossRef]

Felin, Elina, Outi Hälli, Mari Heinonen, Elias Jukola, and Maria Fredriksson-Ahomaa. 2019. Assessment of the Feasibility of Serological Monitoring and On-Farm Information about Health Status for the Future Meat Inspection of Fattening Pigs. *Preventive Veterinary Medicine* 162: 76–82. [CrossRef]

Ferguson, David J. P. 2009. Identification of Faecal Transmission of *Toxoplasma gondii*: Small Science, Large Characters. *International Journal for Parasitology* 39: 871–75. [CrossRef]

Fernandes, Annielle R. F., Diego F. Costa, Muller R. Andrade, Camila S. Bezerra, Rinaldo A. Mota, Clebert J. Alves, Hélio Langoni, and Sérgio S. Azevedo. 2018. Soropositividade e fatores de risco para leptospirose, toxoplasmose e neosporose na população canina do Estado da Paraíba. *Pesquisa Veterinária Brasileira* 38: 957–66. [CrossRef]

Ferreira, Susana Carolina Martins, Francesca Torelli, Sandra Klein, Robert Fyumagwa, William B. Karesh, Heribert Hofer, Frank Seeber, and Marion L. East. 2019. Evidence of High Exposure to *Toxoplasma gondii* in Free-Ranging and Captive African Carnivores. *International Journal for Parasitology. Parasites and Wildlife* 8: 111–17. [CrossRef]

Ferreira-da-Silva, Marialice da Fonseca, Anna C. Takács, Helene S. Barbosa, Uwe Gross, and Carsten G. K. Lüder. 2009. Primary Skeletal Muscle Cells Trigger Spontaneous *Toxoplasma gondii* Tachyzoite-to-Bradyzoite Conversion at Higher Rates than Fibroblasts. *International Journal of Medical Microbiology: IJMM* 299: 381–88. [CrossRef]

Fialho, Cristina Germani, and Flavio Antonio Pacheco de Araujo. 2002. Comparison between Indirect Immunofluorescence and Indirect Haemagglutination Techniques for the Detection of Antibodies against *Toxoplasma gondii* in Swine Sera. *Acta Scientiae Veterinariae* 30: 185–89. [CrossRef]

Figliuolo, Leticia Pinto C., Nobuko Kasai, Alessandra Mara Alves Ragozo, V. S. O. de Paula, Ricardo Augusto Dias, Silvio Luis Pereira De Souza, and Solange Maria Gennari. 2004. Prevalence of Anti-*Toxoplasma gondii* and Anti-Neospora Caninum Antibodies in Ovine from São Paulo State, Brazil. *Veterinary Parasitology* 123: 161–66. [CrossRef]

Figueiredo, Josely F., Deise A. Silva, Dagmar Diniz Cabral, and Jose Roberto Mineo. 2001. Seroprevalence of *Toxoplasma gondii* Infection in Goats by the Indirect Haemagglutination, Immunofluorescence and Immunoenzymatic Tests in the Region of Uberlândia, Brazil. *Memorias Do Instituto Oswaldo Cruz* 96: 687–92. [CrossRef]

Figueredo, Luciana Aguiar, Filipe Dantas-Torres, Eduardo Bento de Faria, Luis Fernando Pita Gondim, Lucilene Simões-Mattos, Sinval Pinto Brandão-Filho, and Rinaldo Aparecido Mota. 2008. Occurrence of Antibodies to Neospora Caninum and *Toxoplasma gondii* in Dogs from Pernambuco, Northeast Brazil. *Veterinary Parasitology* 157: 9–13. [CrossRef] [PubMed]

Figueroa-Castillo, J. Antonio, V. Duarte-Rosas, M. Juárez-Acevedo, Hector Luna-Pastén, and Dolores Correa. 2006. Prevalence of *Toxoplasma gondii* Antibodies in Rabbits (Oryctolagus Cuniculus) from Mexico. *The Journal of Parasitology* 92: 394–95. [CrossRef] [PubMed]

Finger, Mariane Angélica, Eliana Monteforte Cassaro Villalobos, Maria do Carmo Custódio de Souza Hunold Lara, Elenice Maria Sequetin Cunha, Ivan Roque de Barros Filho, Ivan Deconto, Peterson Triches Dornbusch, Leila Sabrina Ullmann, and Alexander Welker Biondo. 2013. Detection of Anti-*Toxoplasma gondii* Antibodies in Carthorses in the Metropolitan Region of Curitiba, Paraná, Brazil. *Revista Brasileira De Parasitologia Veterinaria = Brazilian Journal of Veterinary Parasitology: Orgao Oficial Do Colegio Brasileiro De Parasitologia Veterinaria* 22: 179–81. [CrossRef] [PubMed]

Flegr, Jaroslav, Jan Havlícek, Petr Kodym, Marek Malý, and Zbynek Smahel. 2002. Increased Risk of Traffic Accidents in Subjects with Latent Toxoplasmosis: A Retrospective Case-Control Study. *BMC Infectious Diseases* 2: 11. [CrossRef] [PubMed]

Flegr, Jaroslav, Jirí Klose, Martina Novotná, Miroslava Berenreitterová, and Jan Havlícek. 2009. Increased Incidence of Traffic Accidents in Toxoplasma-Infected Military Drivers and Protective Effect RhD Molecule Revealed by a Large-Scale Prospective Cohort Study. *BMC Infectious Diseases* 9: 72. [CrossRef]

Flegr, Jaroslav, Marek Preiss, Jirí Klose, Jan Havlícek, Martina Vitáková, and Petr Kodym. 2003. Decreased Level of Psychobiological Factor Novelty Seeking and Lower Intelligence in Men Latently Infected with the Protozoan Parasite *Toxoplasma gondii* Dopamine, a Missing Link between Schizophrenia and Toxoplasmosis? *Biological Psychology* 63: 253–68. [CrossRef]

Flegr, Jaroslav. 2007. Effects of Toxoplasma on Human Behavior. *Schizophrenia Bulletin* 33: 757–60. [CrossRef]

Forman, Dan, Nathan West, Janet Francis, and Edward Guy. 2009. The Sero-Prevalence of *Toxoplasma gondii* in British Marine Mammals. *Memorias Do Instituto Oswaldo Cruz* 104: 296–98. [CrossRef]

Fornazari, Felipe, Carlos Roberto Teixeira, Rodrigo Costa da Silva, Maristela Leiva, Silvio César de Almeida, and Helio Langoni. 2011. Prevalence of Antibodies against *Toxoplasma gondii* among Brazilian White-Eared Opossums (Didelphis Albiventris). *Veterinary Parasitology* 179: 238–41. [CrossRef]

Fornazari, Felipe, Helio Langoni, Rodrigo Costa da Silva, Alessandro Guazzelli, Márcio Garcia Ribeiro, and Simone Biagio Chiacchio. 2009. *Toxoplasma gondii* Infection in Wild Boars (Sus Scrofa) Bred in Brazil. *Veterinary Parasitology* 164: 333–34. [CrossRef] [PubMed]

Frank, Rodney K. 2001. An Outbreak of Toxoplasmosis in Farmed Mink (Mustela Vison S.). *Journal of Veterinary Diagnostic Investigation: Official Publication of the American Association of Veterinary Laboratory Diagnosticians, Inc.* 13: 245–49. [CrossRef] [PubMed]

Frazão-Teixeira, Edwards, Natarajan Sundar, Jitender P. Dubey, Michael E. Grigg, and Francisco Carlos Rodrigues de Oliveira. 2011. Multi-Locus DNA Sequencing of *Toxoplasma gondii* Isolated from Brazilian Pigs Identifies Genetically Divergent Strains. *Veterinary Parasitology* 175: 33–39. [CrossRef] [PubMed]

Freytag, H. W., and H. Haas. 1979. [Psychiatric aspects of acquired toxoplasmosis. A case report (author's transl)]. *Der Nervenarzt* 50: 128–31.

Fuentes, Isabel, Mercedes Rodriguez, Carlos J. Domingo, Fernando del Castillo, Teresa Juncosa, and Jorge Alvar. 1996. Urine Sample Used for Congenital Toxoplasmosis Diagnosis by PCR. *Journal of Clinical Microbiology* 34: 2368–71. [CrossRef]

Furtado, Mariana Malzoni, Solange Maria Gennari, Cassia Yumi Ikuta, Anah Tereza de Almeida Jácomo, Zenaide Maria de Morais, Hilda Fátima de Jesus Pena, Grasiela Edith de Oliveira Porfírio, Leandro Silveira, Rahel Sollmann, Gisele Oliveira de Souza, and et al. 2015. Serosurvey of Smooth Brucella, Leptospira Spp. and *Toxoplasma gondii* in Free-Ranging Jaguars (Panthera Onca) and Domestic Animals from Brazil. *PLoS ONE* 10: e0143816. [CrossRef]

Furuta, Takahisa, Yumi Une, Mizuho Omura, Noriko Matsutani, Yasuo Nomura, Takane Kikuchi, Shosaku Hattori, and Yasuhiro Yoshikawa. 2001. Horizontal Transmission of *Toxoplasma gondii* in Squirrel Monkeys (Saimiri Sciureus). *Experimental Animals* 50: 299–306. [CrossRef]

Fusco, Giovanna, Laura Rinaldi, Achille Guarino, Yolande Therese Rose Proroga, Antonella Pesce, De Marco Giuseppina, and Giuseppe Cringoli. 2007. *Toxoplasma gondii* in Sheep from the Campania Region (Italy). *Veterinary Parasitology* 149: 271–74. [CrossRef]

Gajadhar, Alvin A., Jeffery J. Aramini, Greg Tiffin, and Jean-Robert Bisaillon. 1998. Prevalence of *Toxoplasma gondii* in Canadian Market-Age Pigs. *The Journal of Parasitology* 84: 759–63. [CrossRef]

Gajadhar, Alvin A., Lena Measures, Lorry B. Forbes, Christian Kapel, and Jitender P. Dubey. 2004. Experimental *Toxoplasma gondii* Infection in Grey Seals (Halichoerus Grypus). *The Journal of Parasitology* 90: 255–59. [CrossRef]

Gamarra, Jose A., Oscar Cabezón, Marcela Pabón, Maria Cruz Arnal, Daniel F. Luco, Jitender P. Dubey, Christian Gortázar, and Sonia Almeria. 2008. Prevalence of Antibodies against *Toxoplasma gondii* in Roe Deer from Spain. *Veterinary Parasitology* 153: 152–56. [CrossRef] [PubMed]

Gamble, H. Ray, Jitender P. Dubey, and D. N. Lambillotte. 2005. Comparison of a Commercial ELISA with the Modified Agglutination Test for Detection of Toxoplasma Infection in the Domestic Pig. *Veterinary Parasitology* 128: 177–81. [CrossRef] [PubMed]

Gamble, H. Ray, R. C. Brady, and Jitender P. Dubey. 1999. Prevalence of *Toxoplasma gondii* Infection in Domestic Pigs in the New England States. *Veterinary Parasitology* 82: 129–36. [CrossRef] [PubMed]

Garcia, João Luis, Italmar Teodorico Navarro, Liza Ogawa, and Rosângela Claret de Oliveira. 1999. Soroprevalência do toxoplasma gondii, em suínos, bovinos, ovinos e eqüinos, e sua correlação com humanos, felinos e caninos, oriundos de propriedades rurais do norte do Paraná-Brasil. *Ciência Rural* 29: 91–97. [CrossRef]

Garcia, João Luis, Italmar Teodorico Navarro, Luciane Biazzono, Roberta Lemos Freire, José da Silva Guimarães Junior, Andréas L. Cryssafidis, Felipe Monteiro Bugni, Ivo Alexandre Leme da Cunha, Fernando N. Hamada, and Renata Cristina Ferreira Dias. 2007. Protective Activity against Oocyst Shedding in Cats Vaccinated with Crude Rhoptry Proteins of the *Toxoplasma gondii* by the Intranasal Route. *Veterinary Parasitology* 145: 197–206. [CrossRef] [PubMed]

Garcia, João Luis, Italmar Teodorico Navarro, Odilon Vidotto, Solange Maria Gennari, Rosângela Zacarias Machado, Ademir Benedito da Luz Pereira, and Idercio Luiz Sinhorini. 2006a. *Toxoplasma gondii*: Comparison of a Rhoptry-ELISA with IFAT and MAT for Antibody Detection in Sera of Experimentally Infected Pigs. *Experimental Parasitology* 113: 100–105. [CrossRef] [PubMed]

Garcia, João Luis, Solange Maria Gennari, Italmar Teodorico Navarro, Rosângela Zacarias Machado, Selwyn Arligton Headley, Odilon Vidotto, José da Silva Guimarães Junior, Felipe Monteiro Bugni, and Michelle Igarashi. 2008. Evaluation of IFA, MAT, ELISAs and Immunoblotting for the Detection of Anti-*Toxoplasma gondii* Antibodies in Paired Serum and Aqueous Humour Samples from Experimentally Infected Pigs. *Research in Veterinary Science* 84: 237–42. [CrossRef]

Garcia, João Luis, Solange Maria Gennari, Rosângela Zacarias Machado, and Italmar Teodorico Navarro. 2006b. *Toxoplasma gondii*: Detection by Mouse Bioassay, Histopathology, and Polymerase Chain Reaction in Tissues from Experimentally Infected Pigs. *Experimental Parasitology* 113: 267–71. [CrossRef]

Garcia, João Luis, Walfrido Kühl Svoboda, Andréas Lazaros Chryssafidis, Luciano de Souza Malanski, Marcos Massaaki Shiozawa, Lucas de Moraes Aguiar, Gustavo Monteiro Teixeira, Gabriela Ludwig, Lineu Roberto da Silva, Gustavo Monteiro Teixeira, and et al. 2005. Sero-Epidemiological Survey for Toxoplasmosis in Wild New World Monkeys (Cebus Spp.; Alouatta Caraya) at the Paraná River Basin, Paraná State, Brazil. *Veterinary Parasitology* 133: 307–11. [CrossRef]

García-Bocanegra, Ignacio, Jitender P. Dubey, Fernando Martínez, Astrid Vargas, Oscar Cabezón, Irene Zorrilla, Antonio Arenas, and Sonia Almería. 2010a. Factors Affecting Seroprevalence of *Toxoplasma gondii* in the Endangered Iberian Lynx (Lynx Pardinus). *Veterinary Parasitology* 167: 36–42. [CrossRef]

García-Bocanegra, Ignacio, Jitender P. Dubey, Meritxell Simon-Grifé, Oscar Cabezón, Jordi Casal, Alberto Allepuz, Sebastian Napp, and Sonia Almería. 2010b. Seroprevalence and Risk Factors Associated with *Toxoplasma gondii* Infection in Pig Farms from Catalonia, North-Eastern Spain. *Research in Veterinary Science* 89: 85–87. [CrossRef] [PubMed]

García-Bocanegra, Ignacio, Meritxell Simon-Grifé, Marina Sibila, Jitender P. Dubey, Oscar Cabezón, Gerard Martín, and Sonia Almería. 2010c. Duration of Maternally Derived Antibodies in *Toxoplasma gondii* Naturally Infected Piglets. *Veterinary Parasitology* 170: 134–36. [CrossRef] [PubMed]

García-Bocanegra, Ignacio, Oscar Cabezón, Antonio Arenas-Montes, Alfonso Carbonero, Jitender P. Dubey, Anselmo Perea, and Sonia Almería. 2012. Seroprevalence of *Toxoplasma gondii* in Equids from Southern Spain. *Parasitology International* 61: 421–24. [CrossRef]

García-Bocanegra, Ignacio, Oscar Cabezón, E. Hernández, Maria S. Martínez-Cruz, Á Martínez-Moreno, and Francisco Javier Martínez-Moreno. 2013. *Toxoplasma gondii* in Ruminant Species (Cattle, Sheep, and Goats) from Southern Spain. *The Journal of Parasitology* 99: 438–40. [CrossRef]

García-Bocanegra, Ignacio, Meritxell Simon-Grifé, Jitender P. Dubey, Jordi Casal, Gerard E. Martín, Oscar Cabezón, Anselmo Perea, and Sonia Almería. 2010d. Seroprevalence and Risk Factors Associated with *Toxoplasma gondii* in Domestic Pigs from Spain. *Parasitology International* 59: 421–26. [CrossRef] [PubMed]

Gardner, Ian A., Henrik Stryhn, Peter Lind, and Michael T. Collins. 2000. Conditional Dependence between Tests Affects the Diagnosis and Surveillance of Animal Diseases. *Preventive Veterinary Medicine* 45: 107–22. [CrossRef]

Gaskell, Elizabeth A., Judith E. Smith, John W. Pinney, Dave R. Westhead, and Glenn A. McConkey. 2009. A Unique Dual Activity Amino Acid Hydroxylase in *Toxoplasma gondii*. *PLoS ONE* 4: e4801. [CrossRef] [PubMed]

Gatkowska, Justyna, Elzbieta Hiszczynska-Sawicka, Jozef Kur, Lucyna Holec, and Henryka Dlugonska. 2006. *Toxoplasma gondii*: An Evaluation of Diagnostic Value of Recombinant Antigens in a Murine Model. *Experimental Parasitology* 114: 220–27. [CrossRef]

Gauss, Ceres B. L., Jitender P. Dubey, Dolo Vidal, Francisco Ruiz, Joaquin Vicente, Ignasi Marco, Santiago Lavin, Christian Gortazar, and Sonia Almería. 2005. Seroprevalence of *Toxoplasma gondii* in Wild Pigs (Sus Scrofa) from Spain. *Veterinary Parasitology* 131: 151–56. [CrossRef]

Gauss, Ceres B. L., Jitender P. Dubey, Dolo Vidal, Oscar Cabezón, Francisco Ruiz-Fons, Joaquin Vicente, Ignasi Marco, Santiago Lavin, Christian Gortazar, and Sonia Almería. 2006. Prevalence of *Toxoplasma gondii* Antibodies in Red Deer (Cervus Elaphus) and Other Wild Ruminants from Spain. *Veterinary Parasitology* 136: 193–200. [CrossRef]

Gauss, Ceres B. L., Sonia Almería, A. Ortuño, F. Garcia, and Jitender P. Dubey. 2003. Seroprevalence of *Toxoplasma gondii* Antibodies in Domestic Cats from Barcelona, Spain. *The Journal of Parasitology* 89: 1067–68. [CrossRef] [PubMed]

Gazeta, Gilberto Salles, A.E.A. Dutra, and Antonio N. Norberg. 1997. Frequencia de Anticorpos Anti-*Toxoplasma gondii* Em Soros de Equinos No Estado Do Rio de Janeiro, Brasil. *Revista Brasileira De Parasitologia Veterinaria* 6: 87–91.

Gazzonis, Alessia Libera, Fabrizia Veronesi, Anna Rita Di Cerbo, Sergio Aurelio Zanzani, Giulia Molineri, Iolanda Moretta, Annabella Moretti, Daniela Piergili Fioretti, Anna Invernizzi, and Maria Teresa Manfredi. 2015. *Toxoplasma gondii* in Small Ruminants in Northern Italy—Prevalence and Risk Factors. *Annals of Agricultural and Environmental Medicine: AAEM* 22: 62–68. [CrossRef]

Gazzonis, Alessia Libera, Sergio Aurelio Zanzani, Luca Villa, and Maria Teresa Manfredi. 2020. *Toxoplasma gondii* Infection in Meat-Producing Small Ruminants: Meat Juice Serology and Genotyping. *Parasitology International* 76: 102060. [CrossRef]

Gebremedhin, Endrias Zewdu, and Daniel Gizaw. 2014. Seroprevalence of *Toxoplasma gondii* Infection in Sheep and Goats in Three Districts of Southern Nations, Nationalities and Peoples' Region of Ethiopia. *World Applied Sciences Journal* 31: 1891–96. [CrossRef]

Gebremedhin, Endrias Zewdu, Mukarim Abdurahaman, Tsehaye Hadush, and Tesfaye Sisay Tessema. 2014. Seroprevalence and Risk Factors of *Toxoplasma gondii* Infection in Sheep and Goats Slaughtered for Human Consumption in Central Ethiopia. *BMC Research Notes* 7: 696. [CrossRef]

Gencay, Yilmaz Emre, Kader Yildiz, Sami Gokpinar, and Abdullah Leblebicier. 2013. A Potential Infection Source for Humans: Frozen Buffalo Meat Can Harbour Tissue Cysts of *Toxoplasma gondii*. *Food Control* 30: 86–89. [CrossRef]

Gerhold, Richard W., Elizabeth W. Howerth, and David S. Lindsay. 2005. Sarcocystis Neurona-Associated Meningoencephalitis and Description of Intramuscular Sarcocysts in a Fisher (Martes Pennanti). *Journal of Wildlife Diseases* 41: 224–30. [CrossRef]

Gerhold, Richard W., Pooja Saraf, Alycia Chapman, Xuan Zou, Graham Hickling, William H. Stiver, Allan Houston, Marcy Souza, and Chunlei Su. 2017. *Toxoplasma gondii* Seroprevalence and Genotype Diversity in Select Wildlife Species from the Southeastern United States. *Parasites & Vectors* 10: 508. [CrossRef]

Ghazaei, Ciamak. 2006. Serological Survey of Antibodies to *Toxoplasma gondii*. *African Journal of Health Sciences* 13: 131–34. [CrossRef]

Ghazy, Alaa A., Raafat M. Shaapan, and Eman H. Abdel-Rahman. 2007. Comparative Serological Diagnosis of Toxoplasmosis in Horses Using Locally Isolated *Toxoplasma gondii*. *Veterinary Parasitology* 145: 31–36. [CrossRef]

Gilot-Fromont, Emmanuelle, Dominique Aubert, S. Belkilani, P. Hermitte, O. Gibout, Regine Geers, and Isabelle Villena. 2009. Landscape, Herd Management and within-Herd Seroprevalence of *Toxoplasma gondii* in Beef Cattle Herds from Champagne-Ardenne, France. *Veterinary Parasitology* 161: 36–40. [CrossRef]

Giraldi, J.H. 2004. Serology and Histopathology of *Toxoplasma gondii* and Neospora Caninum in Dogs with Neurological Disorders. *Veterinary Bulletin* 74: 216.

Gisbert Algaba, Ignacio, Bavo Verhaegen, Jean Benjamin Murat, Wim Coucke, Aurélien Mercier, Eric Cox, Pierre Dorny, Katelijne Dierick, and Stephane De Craeye. 2020. Molecular Study of *Toxoplasma gondii* Isolates Originating from Humans and Organic Pigs in Belgium. *Foodborne Pathogens and Disease* 17: 316–21. [CrossRef] [PubMed]

Glor, Sabine B., Renate Edelhofer, Felix Grimm, Peter Deplazes, and Walter Basso. 2013. Evaluation of a Commercial ELISA Kit for Detection of Antibodies against *Toxoplasma gondii* in Serum, Plasma and Meat Juice from Experimentally and Naturally Infected Sheep. *Parasites & Vectors* 6: 85. [CrossRef]

Glynn, M. Kathleen. 2008. Handbook of Zoonoses: Identification and Prevention. *Emerging Infectious Diseases* 14: 354. [CrossRef]

Goez, Yaşar, C. Babuer, A. Aydin, and Sinan Kilic. 2007. Seroprevalence of Toxoplasmosis, Brucellosis and Listeriosis in Horses in Hakkari, Eastern Region of Turkey. *Revue De Medecine Veterinaire* 158: 11.

Golkar, Majid, Kayhan Azadmanesh, Ghader Khalili, Baharak Khoshkholgh-Sima, Jalal Babaie, Corinne Mercier, Marie-Pierre Brenier-Pinchart, Hélène Fricker-Hidalgo, Hervé Pelloux, and Marie-France Cesbron-Delauw. 2008. Serodiagnosis of Recently Acquired *Toxoplasma gondii* Infection in Pregnant Women Using Enzyme-Linked Immunosorbent Assays with a Recombinant Dense Granule GRA6 Protein. *Diagnostic Microbiology and Infectious Disease* 61: 31–39. [CrossRef]

Gondim, Leane S. Q., Kiyoko Abe-Sandes, Rosângela S. Uzêda, Mariana S. A. Silva, Sara L. Santos, Rinaldo A. Mota, Sineide M. O. Vilela, and Luis F. P. Gondim. 2010. *Toxoplasma gondii* and Neospora Caninum in Sparrows (Passer Domesticus) in the Northeast of Brazil. *Veterinary Parasitology* 168: 121–24. [CrossRef]

Górecki, Marcin T., Izabela Andrzejewska, and Ryszard Steppa. 2008. Is Order of Voluntarily Entrance to Milking Parlour Related to *Toxoplasma gondii* Infection in Sheep—A Brief Note. *Applied Animal Behaviour Science* 110: 392–96. [CrossRef]

Gorman, Texia, J. P. Arancibia, Myriam Lorca, D. Hird, and Hector Alcaino. 1999. Seroprevalence of *Toxoplasma gondii* Infection in Sheep and Alpacas (Llama Pacos) in Chile. *Preventive Veterinary Medicine* 40: 143–49. [CrossRef]

Goz, Y., A. Aydin, and E. Ceylan. 2006. Seroprevalence of *Toxoplasma gondii* Infection in Goats. *Indian Veterinary Journal* 82: 11–13.

Gross, Uwe, W. Bohne, J. Schröder, Thomas Roos, and J. Heesemann. 1993. Comparison of a Commercial Enzyme Immunoassay and an Immunoblot Technique for Detection of Immunoglobulin a Antibodies to *Toxoplasma gondii*. *European Journal of Clinical Microbiology & Infectious Diseases: Official Publication of the European Society of Clinical Microbiology* 12: 636–39. [CrossRef]

Güçlü, Zeynep, Zafer Karaer, Cahit Babür, and Selçuk Kiliç. 2007. Investigation of *Toxoplasma gondii* Antibodies in Sport Horses Bred in Ankara Province. *Turkiye Parazitolojii Dergisi* 31: 264–67. [PubMed]

Guerra, Neurisvan R., Jonatas C. Almeida, Elâine L. Silva, Edson M. Silva, José A. M. Santos, Raphael Lepold, Rinaldo A. Mota, and Leucio C. Alves. 2018. Soroprevalência de Toxoplasma gondii em equídeos do Nordeste do Brasil. *Pesquisa Veterinária Brasileira* 38: 400–406. [CrossRef]

Gulinello, Maria, Mariana Acquarone, John H. Kim, David C. Spray, Helene S. Barbosa, Rani Sellers, Herbert B. Tanowitz, and Louis M. Weiss. 2010. Acquired Infection with *Toxoplasma gondii* in Adult Mice Results in Sensorimotor Deficits but Normal Cognitive Behavior despite Widespread Brain Pathology. *Microbes and Infection* 12: 528–37. [CrossRef]

Gupta, G. D., Jeffrey Lakritz, Jae-Hoon Kim, Dae-Yong Kim, Jin-Kap Kim, and Antoinette E. Marsh. 2002. Seroprevalence of Neospora, *Toxoplasma gondii* and Sarcocystis Neurona Antibodies in Horses from Jeju Island, South Korea. *Veterinary Parasitology* 106: 193–201. [CrossRef]

Gustafsson, Katarina, Arvid Uggla, and Bertil Järplid. 1997a. *Toxoplasma gondii* Infection in the Mountain Hare (Lepus Timidus) and Domestic Rabbit (Oryctolagus Cuniculus). I. Pathology. *Journal of Comparative Pathology* 117: 351–60. [CrossRef]

Gustafsson, Katarina, Eva Wattrang, Caroline Fossum, Peter M. Heegaard, Peter Lind, and Arvid Uggla. 1997b. *Toxoplasma gondii* Infection in the Mountain Hare (Lepus Timidus) and Domestic Rabbit (Oryctolagus Cuniculus). II. Early Immune Reactions. *Journal of Comparative Pathology* 117: 361–69. [CrossRef]

Györke, Adriana, Marieke Opsteegh, Viorica Mircean, Anamaria Iovu, and Vasile Cozma. 2011. *Toxoplasma gondii* in Romanian Household Cats: Evaluation of Serological Tests, Epidemiology and Risk Factors. *Preventive Veterinary Medicine* 102: 321–28. [CrossRef] [PubMed]

Hall, Susan M., A. Pandit, Ajit Golwilkar, and T. S. Williams. 1999. How Do Jains Get Toxoplasma Infection? *Lancet (London, England)* 354: 486–87. [CrossRef]

Hancock, Katie, Lori A. Thiele, Anne M. Zajac, Francois Elvingert, and David S. Lindsay. 2005. Prevalence of Antibodies to *Toxoplasma gondii* in Raccoons (Procyon Lotor) from an Urban Area of Northern Virginia. *The Journal of Parasitology* 91: 694–95. [CrossRef] [PubMed]

Harfoush, Mohamed, and Abd El-Naby Tahoon. 2010. Seroprevalence of *Toxoplasma gondii* Antibodies in Domestic Ducks, Free-Range Chickens, Turkeys and Rabbits in Kafr El-Sheikh Governorate Egypt. *Journal of the Egyptian Society of Parasitology* 40: 295–302. [PubMed]

Haridy, Fouad M., Nagla Mostafa K. Saleh, Hazem H. M. Khalil, and Tosson A. Morsy. 2010. Anti-*Toxoplasma gondii* Antibodies in Working Donkeys and Donkey's Milk in Greater Cairo, Egypt. *Journal of the Egyptian Society of Parasitology* 40: 459–64.

Harris, S., and D. B. Jones. 1997. Optimisation of the Polymerase Chain Reaction. *British Journal of Biomedical Science* 54: 166–73.

Hartley, Matt, and Anthony English. 2005. A Seroprevalence Survey of *Toxoplasma gondii* in Common Wombats (Vombatus Ursinus). *European Journal of Wildlife Research* 51: 65–67. [CrossRef]

Havlícek, Jan, Zdenka G. Gasová, Andrew P. Smith, Karel Zvára, and Jaroslav Flegr. 2001. Decrease of Psychomotor Performance in Subjects with Latent "asymptomatic" Toxoplasmosis. *Parasitology* 122, Pt 5: 515–20. [CrossRef]

Hejlicek, Karel, and Ivan Literak. 1994a. Prevalence of Antibodies to *Toxoplasma gondii* in Horses in the Czech Republic. *Acta Parasitologica* 39: 217–19.

Hejlicek, Karel, and Ivan Literak. 1994b. Toxoplazmoza Jatecnych Zvirat. Hygienicka Problematika. *Veterinarstvi* 44: 576–81.

Hendrix, C.M., and Margi Sirois. 2007. Laboratory Procedures for Veterinary Technicians, 7th Edition—9780323595384. Available online: https://evolve.elsevier.com/cs/product/9780323595384?role=student (accessed on 24 March 2023).

Henriksen, Per, Hans Henrik Dietz, A. Uttenthal, and Mogens Hansen. 1994. Seroprevalence of *Toxoplasma gondii* in Danish Farmed Mink (Mustela Vison S.). *Veterinary Parasitology* 53: 1–5. [CrossRef]

Hernández, Marivel, Jaime Gómez-Laguna, Carmen Tarradas, Inmaculada Luque, Rosa García-Valverde, Lucia Reguillo, and Rafael Jesus Astorga. 2014. A Serological Survey of *Brucella* Spp., *Salmonella* Spp., *Toxoplasma gondii* and *Trichinella* Spp. in Iberian Fattening Pigs Reared in Free-Range Systems. *Transboundary and Emerging Diseases* 61: 477–81. [CrossRef]

Herrero, Laura, María Jesús Gracia, Consuelo Pérez-Arquillué, Regina Lázaro, Marta Herrera, Antonio Herrera, and Susana Bayarri. 2016. *Toxoplasma gondii*: Pig Seroprevalence, Associated Risk Factors and Viability in Fresh Pork Meat. *Veterinary Parasitology* 224: 52–59. [CrossRef]

Herrmann, Daland C., Nikola Pantchev, Majda Globokar Vrhovec, Dieter Barutzki, Hendrik Wilking, Andreas Fröhlich, Carsten G. K. Lüder, Franz J. Conraths, and Gereon Schares. 2010. Atypical *Toxoplasma gondii* Genotypes Identified in Oocysts Shed by Cats in Germany. *International Journal for Parasitology* 40: 285–92. [CrossRef]

Hill, Dolores E., Sreekumar Chirukandoth, J. P. Dubey, Joan K. Lunney, and H. Ray Gamble. 2006. Comparison of Detection Methods for *Toxoplasma gondii* in Naturally and Experimentally Infected Swine. *Veterinary Parasitology* 141: 9–17. [CrossRef] [PubMed]

Hill, Nichola J., Jitender P. Dubey, L. Vogelnest, Michelle L. Power, and Elizabeth M. Deane. 2008. Do Free-Ranging Common Brushtail Possums (Trichosurus Vulpecula) Play a Role in the Transmission of *Toxoplasma gondii* within a Zoo Environment? *Veterinary Parasitology* 152: 202–9. [CrossRef] [PubMed]

Hill, R. E., Jeffrey James Zimmerman, Robert W. Wills, Sharon Patton, and William R. Clark. 1998. Seroprevalence of Antibodies against *Toxoplasma gondii* in Free-Ranging Mammals in Iowa. *Journal of Wildlife Diseases* 34: 811–15. [CrossRef]

Holec-Gąsior, Lucyna, Bartosz Dominiak-Górski, and Józef Kur. 2015. First Report of Seroprevalence of *Toxoplasma gondii* Infection in Sheep in Pomerania, Northern Poland. *Annals of Agricultural and Environmental Medicine: AAEM* 22: 604–7. [CrossRef] [PubMed]

Holliman, Richard E. 1997. Toxoplasmosis, Behaviour and Personality. *The Journal of Infection* 35: 105–10. [CrossRef]

Hornok, Sándor, Renate Edelhofer, Anja Joachim, Róbert Farkas, Krisztián Berta, Attila Répási, and Béla Lakatos. 2008. Seroprevalence of *Toxoplasma gondii* and Neospora Caninum Infection of Cats in Hungary. *Acta Veterinaria Hungarica* 56: 81–88. [CrossRef]

Hosseininejad, Morteza, Hamidreza R. Azizi, Farzaneh Hosseini, and Gereon Schares. 2009. Development of an Indirect ELISA Test Using a Purified Tachyzoite Surface Antigen SAG1 for Sero-Diagnosis of Canine *Toxoplasma gondii* Infection. *Veterinary Parasitology* 164: 315–19. [CrossRef] [PubMed]

Hotea, Ionela, Gheorghe Dărăbuş, Marius Ilie, Kalman Imre, Ion Oprescu, Sorin Morariu, and Narcisa Mederle. 2009. The Identification of *Toxoplasma gondii* Infection in Sheep by ELISA from Arad County. *Lucrari Stiintifice—Universitatea de Stiinte Agricole u Burutului Timisoara, Medicina Veterinara* 42: 16–21.

Hotea, Ionela, Marius Ilie, Mirela Imre, Denisa Sorescu, and Gheorghe Dărăbuş. 2011a. Determining of Toxoplasmosis Seroprevalence, by ELISA, in Lambs in Timis County. *Lucrări Ştiinţifice Medicină Veterinară, Iasi* 54: 161–64.

Hotea, Ionela, Marius Ilie, Mirela Imre, Denisa Sorescu, Dinu Indre, Ileana Brudiu, Olimpia Colibar, and Gheorghe Dărăbuş. 2012. *Toxoplasma gondii* Seroprevalence in Cats and Sheep from Caras-Severin County, Romania. *Lucrari Stiintifice-Universitatea de Stiinte Agricole a Banatului Timisoara, Medicina Veterinara* 45: 104–9.

Hotea, Ionela, Ion Oprescu, Marius Ilie, Kalman Imre, Adrian Balint, and Gheorghe Dărăbuş. 2011b. Prevalence of *Toxoplasma gondii* Infection, by ELISA, in Rams in Timis County. *Lucrări Ştiintifice Medicină Veterinară Timisoara* 54: 325–28.

Hou, Zhao-Feng, Shi-Jie Su, Dan-Dan Liu, Le-le Wang, Chuan-Li Jia, Zhen-Xing Zhao, Yi-Fei Ma, Qiao-Qiao Li, Jin-Jun Xu, and Jian-Ping Tao. 2018. Prevalence, Risk Factors and Genetic Characterization of *Toxoplasma gondii* in Sick Pigs and Stray Cats in Jiangsu Province, Eastern China. *Infection, Genetics and Evolution: Journal of Molecular Epidemiology and Evolutionary Genetics in Infectious Diseases* 60: 17–25. [CrossRef] [PubMed]

Hove, Thokozani, and Jitender P. Dubey. 1999. Prevalence of *Toxoplasma gondii* Antibodies in Sera of Domestic Pigs and Some Wild Game Species from Zimbabwe. *The Journal of Parasitology* 85: 372–73. [CrossRef]

Hove, Thokozani, and Samson Mukaratirwa. 2005. Seroprevalence of *Toxoplasma gondii* in Farm-Reared Ostriches and Wild Game Species from Zimbabwe. *Acta Tropica* 94: 49–53. [CrossRef]

Hove, Thokozani, Peter Lind, and Samson Mukaratirwa. 2005a. Seroprevalence of *Toxoplasma gondii* Infection in Goats and Sheep in Zimbabwe. *The Onderstepoort Journal of Veterinary Research* 72: 267–72. [CrossRef]

Hove, Thokozani, Peter Lind, and Samson Mukaratirwa. 2005b. Seroprevalence of *Toxoplasma gondii* Infection in Domestic Pigs Reared under Different Management Systems in Zimbabwe. *The Onderstepoort Journal of Veterinary Research* 72: 231–37. [CrossRef]

Hsu, Vasha, David C. Grant, Anne M. Zajac, Sharon G. Witonsky, and David S. Lindsay. 2011. Prevalence of IgG Antibodies to Encephalitozoon Cuniculi and *Toxoplasma gondii* in Cats with and without Chronic Kidney Disease from Virginia. *Veterinary Parasitology* 176: 23–26. [CrossRef]

Huang, Cuiqin Q., Y. Y. Lin, Ailing L. Dai, X. H. Li, X. Y. Yang, Ziguo G. Yuan, and X. Q. Zhu. 2010. Seroprevalence of *Toxoplasma gondii* Infection in Breeding Sows in Western Fujian Province, China. *Tropical Animal Health and Production* 42: 115–18. [CrossRef] [PubMed]

Hughes, Jacqueline M., Denise Thomasson, Phillip Simon Craig, S. Georgin, A. Pickles, and Geoff Hide. 2008. Neospora Caninum: Detection in Wild Rabbits and Investigation of Co-Infection with *Toxoplasma gondii* by PCR Analysis. *Experimental Parasitology* 120: 255–60. [CrossRef] [PubMed]

Huong, Lam Thi Thu, and Jitender P. Dubey. 2007. Seroprevalence of *Toxoplasma gondii* in Pigs from Vietnam. *The Journal of Parasitology* 93: 951–52. [CrossRef]

Hůrková, Lada, and David Modrý. 2006. PCR Detection of Neospora Caninum, *Toxoplasma gondii* and Encephalitozoon Cuniculi in Brains of Wild Carnivores. *Veterinary Parasitology* 137: 150–54. [CrossRef]

Inci, A., and Z. Karaer. 1996. Detection of anti-toxoplasma gondii antibodies using Sabin-Feldman dye test in horses in military. *Acta Parasitologica Turcica (Turkey)*. Available online: https://scholar.google.com/scholar_lookup?title=Detection+of+anti-toxoplasma+gondii+antibodies+using+Sabin-Feldman+dye+test+in+horses+in+military&author=Inci%2C+A.&publication_year=1996 (accessed on 23 September 2020).

Inoue, Isamu, Chen Sau Leow, Darman Husin, Kayoko Matsuo, and Puguh Darmani. 2001. A Survey of *Toxoplasma gondii* Antibodies in Pigs in Indonesia. *The Southeast Asian Journal of Tropical Medicine and Public Health* 32: 38–40.

Iovu, Anamaria, Adriana Györke, Viorica Mircean, Raluca Gavrea, and Vasile Cozma. 2012. Seroprevalence of *Toxoplasma gondii* and Neospora Caninum in Dairy Goats from Romania. *Veterinary Parasitology* 186: 470–74. [CrossRef]

Irabuena, O., D. Fernández Abella, N. Villegas, L. Collazo, and J. Battistoni. 2006. Incidence of *Toxoplasma gondii* Infection during Pregnancy in Sheep Reproduction. *Producc. Ovina* 17: 61–67.

Isaac-Renton, Judith, William R. Bowie, Arlene King, G. Stewart Irwin, Corrine S. Ong, C. P. Fung, M. Omar Shokeir, and Jitender P. Dubey. 1998. Detection of *Toxoplasma gondii* Oocysts in Drinking Water. *Applied and Environmental Microbiology* 64: 2278–80. [CrossRef]

Ishag, Manal, E. Magzoub, and M. Majid. 2006. Detection of *Toxoplasma gondii* Tachyzoites in the Milk of Experimentally Infected Lactating She-Camels. *Journal of Animal and Veterinary Advances* 5.

Jakubek, Eva-Britt, Anna Lundén, and Arvid Uggla. 2006. Seroprevalences of *Toxoplasma gondii* and Neospora Sp. Infections in Swedish Horses. *Veterinary Parasitology* 138: 194–99. [CrossRef]

Jakubek, Eva-Britt, Robert Farkas, Vilmos Pálfi, and Jens G. Mattsson. 2007. Prevalence of Antibodies against *Toxoplasma gondii* and Neospora Caninum in Hungarian Red Foxes (Vulpes Vulpes). *Veterinary Parasitology* 144: 39–44. [CrossRef] [PubMed]

Jakubek, Eva-Britt, Roland Mattsson, Torsten Mörner, Jens G. Mattsson, and Dolores Gavier-Widén. 2012. Potential Application of Serological Tests on Fluids from Carcasses: Detection of Antibodies against *Toxoplasma gondii* and Sarcoptes Scabiei in Red Foxes (Vulpes Vulpes). *Acta Veterinaria Scandinavica* 54: 13. [CrossRef]

James, Kaitlyn E., Woutrina A. Smith, Andrea E. Packham, Patricia A. Conrad, and Nicola Pusterla. 2017. *Toxoplasma gondii* Seroprevalence and Association with Equine Protozoal Myeloencephalitis: A Case-Control Study of Californian Horses. *Veterinary Journal (London, England: 1997)* 224: 38–43. [CrossRef] [PubMed]

Javadi, S., S. Asri Rezaei, S. Abdollahi, M. Tvasoli, M. Khodaverdizadeh, and N. Delirezh. 2008. *Toxoplasma gondii* Specific Antibodies in Asymptomatic Dogs and Cats | IVIS. European Veterinary Conference. Voorjaarsdagen 258. April 26. Available online: https://www.ivis.org/library/evc/evc-voorjaarsdagen-amsterdam-2008/toxoplasma-gondii-specific-antibodies-asymptomatic-dogs-and-cats (accessed on 18 November 2021).

Jay, James M., Martin J. Loessner, and David A. Golden. 2005. *Modern Food Microbiology*, 7th ed. New York: Springer.

Jeklova, Edita, Vladimir Jekl, Kamil Kovarcik, Karel Hauptman, Bretislav Koudela, Helena Neumayerova, Zdenek Knotek, and Martin Faldyna. 2010. Usefulness of Detection of Specific IgM and IgG Antibodies for Diagnosis of Clinical Encephalitozoonosis in Pet Rabbits. *Veterinary Parasitology* 170: 143–48. [CrossRef] [PubMed]

Jiang, Tao, Dachun Gong, Li-An Ma, Hao Nie, Yanqin Zhou, Baoan Yao, and Junlong Zhao. 2008. Evaluation of a Recombinant MIC3 Based Latex Agglutination Test for the Rapid Serodiagnosis of *Toxoplasma gondii* Infection in Swines. *Veterinary Parasitology* 158: 51–56. [CrossRef]

Jiménez-Martín, Débora, Ignacio García-Bocanegra, Sonia Almería, Sabrina Castro-Scholten, Jitender P. Dubey, Manuel A. Amaro-López, and David Cano-Terriza. 2020. Epidemiological Surveillance of *Toxoplasma gondii* in Small Ruminants in Southern Spain. *Preventive Veterinary Medicine* 183: 105137. [CrossRef]

Jittapalapong, Sathaporn, Arkom Sangvaranond, Nongnuch Pinyopanuwat, Wissanuwat Chimnoi, Witaya Khachaeram, Seiichi Koizumi, and Soichi Maruyama. 2005. Seroprevalence of *Toxoplasma gondii* Infection in Domestic Goats in Satun Province, Thailand. *Veterinary Parasitology* 127: 17–22. [CrossRef]

Jittapalapong, Sathaporn, Burin Nimsupan, Nongnuch Pinyopanuwat, Wissanuwat Chimnoi, Hidenori Kabeya, and Soichi Maruyama. 2007. Seroprevalence of *Toxoplasma gondii* Antibodies in Stray Cats and Dogs in the Bangkok Metropolitan Area, Thailand. *Veterinary Parasitology* 145: 138–41. [CrossRef]

Johnson, Alan M. 1997. Speculation on Possible Life Cycles for the Clonal Lineages in the Genus Toxoplasma. *Parasitology Today (Personal Ed.)* 13: 393–97. [CrossRef]

Johnson, Alan M., Peter Phillips, and David Jenkins. 1990. Prevalence of *Toxoplasma gondii* Antibodies in Dingoes. *Journal of Wildlife Diseases* 26: 383–86. [CrossRef]

Johnson, Alan M. 1996. Australian Native Marsupials as Vector for Toxoplasmosis. *Abstracts of the XIVth Internat. Congr. For Tropical Med. and Malaria. Nagasaki, Japan.* 14: 17–22.

Jokelainen, Pikka, Anu Näreaho, Outi Hälli, Mari Heinonen, and Antti Sukura. 2012a. Farmed Wild Boars Exposed to *Toxoplasma gondii* and *Trichinella* Spp. *Veterinary Parasitology* 187: 323–27. [CrossRef] [PubMed]

Jokelainen, Pikka, Anu Näreaho, Suvi Knaapi, Antti Oksanen, Ulla Rikula, and Antti Sukura. 2010. *Toxoplasma gondii* in Wild Cervids and Sheep in Finland: North-South Gradient in Seroprevalence. *Veterinary Parasitology* 171: 331–36. [CrossRef] [PubMed]

Jokelainen, Pikka, Maarja Tagel, Kerli Mõtus, Arvo Viltrop, and Brian Lassen. 2017. *Toxoplasma gondii* Seroprevalence in Dairy and Beef Cattle: Large-Scale Epidemiological Study in Estonia. *Veterinary Parasitology* 236: 137–43. [CrossRef] [PubMed]

Jokelainen, Pikka, Outi Simola, Elina Rantanen, Anu Näreaho, Hannes Lohi, and Antti Sukura. 2012b. Feline Toxoplasmosis in Finland: Cross-Sectional Epidemiological Study and Case Series Study. *Journal of Veterinary Diagnostic Investigation: Official Publication of the American Association of Veterinary Laboratory Diagnosticians, Inc.* 24: 1115–24. [CrossRef] [PubMed]

Jones, Jeffrey L., Deanna Kruszon-Moran, Kolby Sanders-Lewis, and Marianna Wilson. 2007. *Toxoplasma gondii* Infection in the United States, 1999 2004, Decline from the Prior Decade. *The American Journal of Tropical Medicine and Hygiene* 77: 405–10. [CrossRef]

Jordan, Carly N., Taranjit Kaur, Kiana Koenen, Stephen DeStefano, Anne M. Zajac, and David S. Lindsay. 2005. Prevalence of Agglutinating Antibodies to *Toxoplasma gondii* and Sarcocystis Neurona in Beavers (Castor Canadensis) from Massachusetts. *The Journal of Parasitology* 91: 1228–29. [CrossRef]

Kajerová, Vilma, Ivan Literák, Eva Bártová, and Kamil Sedlák. 2003. Experimental Infection of Budgerigars (Melopsittacus Undulatus) with a Low Virulent K21 Strain of *Toxoplasma gondii. Veterinary Parasitology* 116: 297–304. [CrossRef]

Kamani, Joshua, Aliyu U. Mani, Hussaini A. Kumshe, James P. Yidawi, and Godwin O. Egwu. 2010. Prevalence of *Toxoplasma gondii* Antibodies in Cats in Maiduguri, Northeastern Nigeria. *Acta Parasitologica* 55: 94–95. [CrossRef]

Kannan, Geetha, Krisztina Moldovan, Jian-Chun Xiao, Robert H. Yolken, Lorraine Jones-Brando, and Mikhail V. Pletnikov. 2010. *Toxoplasma gondii* Strain-Dependent Effects on Mouse Behaviour. *Folia Parasitologica* 57: 151–55. [CrossRef]

Karatepe, Bilge, Cahit Babur, M. Karatepe, A. Cakmak, and S. Nalbantoglu. 2004. Seroprevalence of *Toxoplasma gondii* in Cattle in Nigde Province. *Veterinary Bulletin* 74: 1181.

Karatepe, Bilge, Cahit Babür, M. Karatepe, and Selçuk Kiliç. 2010. Seroprevalence of Toxoplasmosis in Horses in Niğde Province of Turkey. *Tropical Animal Health and Production* 42: 385–89. [CrossRef] [PubMed]

Kasper, David C., Kambis Sadeghi, Andrea-Romana Prusa, Georg H. Reischer, Klaus Kratochwill, Elisabeth Förster-Waldl, Nicole Gerstl, Michael Hayde, Arnold Pollak, and Kurt R. Herkner. 2009. Quantitative Real-Time Polymerase Chain Reaction for the Accurate Detection of *Toxoplasma gondii* in Amniotic Fluid. *Diagnostic Microbiology and Infectious Disease* 63: 10–15. [CrossRef] [PubMed]

Kelly, Terra R., and Jonathan M. Sleeman. 2003. Morbidity and Mortality of Red Foxes (Vulpes Vulpes) and Gray Foxes (Urocyon Cinereoargenteus) Admitted to the Wildlife Center of Virginia, 1993-2001. *Journal of Wildlife Diseases* 39: 467–69. [CrossRef] [PubMed]

Khodaverdi, Majid, and Gholamreza Razmi. 2019. A Serological and Parasitological Study of *Toxoplasma gondii* Infection in Stray Cats of Mashhad, Khorasan Razavi Province, Iran. *Veterinary Research Forum: An International Quarterly Journal* 10: 119–23. [CrossRef]

Kijlstra, Aize, Bastiaan Meerburg, Jan Cornelissen, Stéphane De Craeye, Pieter Vereijken, and Erik Jongert. 2008. The Role of Rodents and Shrews in the Transmission of *Toxoplasma gondii* to Pigs. *Veterinary Parasitology* 156: 183–90. [CrossRef] [PubMed]

Klein, Stephanie Kristin, Michael Wendt, Wolfgang Baumgärtner, and Peter Wohlsein. 2010. Systemic Toxoplasmosis and Concurrent Porcine Circovirus-2 Infection in a Pig. *Journal of Comparative Pathology* 142: 228–34. [CrossRef]

Klun, Ivana, Olgica Djurković-Djaković, and Philippe Thulliez. 2007. Comparison of a Commercial ELISA with the Modified Agglutination Test for the Detection of *Toxoplasma gondii* Infection in Naturally Exposed Sheep. *Zoonoses and Public Health* 54: 165–68. [CrossRef] [PubMed]

Klun, Ivana, Aleksandra Uzelac, Isabelle Villena, Aurélien Mercier, Branko Bobić, Aleksandra Nikolić, and Irena Rajnpreht. 2017. The First Isolation and Molecular Characterization of *Toxoplasma gondii* from Horses in Serbia. *Parasites & Vectors* 10: 167. [CrossRef]

Klun, Ivana, Olgica Djurković-Djaković, Sofija Katić-Radivojević, and Aleksandra Nikolić. 2006. Cross-Sectional Survey on *Toxoplasma gondii* Infection in Cattle, Sheep and Pigs in Serbia: Seroprevalence and Risk Factors. *Veterinary Parasitology* 135: 121–31. [CrossRef]

Kocazeybek, Bekir, Yasar Ali Oner, Recep Turksoy, Cahit Babur, Huseyin Cakan, Nilgun Sahip, Ali Unal, Abdi Ozaslan, Selcuk Kilic, Suat Saribas, and et al. 2009. Higher Prevalence of Toxoplasmosis in Victims of Traffic Accidents Suggest Increased Risk of Traffic Accident in Toxoplasma-Infected Inhabitants of Istanbul and Its Suburbs. *Forensic Science International* 187: 103–8. [CrossRef]

Kocher, Thomas David, A. C. Wilson, and R. A. Gibbs. 1990. DNA Amplification by the Polymerase Chain Reaction. *Analytical Chemistry* 62: 1202–14. [CrossRef]

Kofoed, Kristina Grønbech, Mia Vorslund-Kiær, Henrik Vedel Nielsen, Lis Alban, and Maria Vang Johansen. 2017. Sero-Prevalence of *Toxoplasma gondii* in Danish Pigs. *Veterinary Parasitology, Regional Studies and Reports* 10: 136–38. [CrossRef] [PubMed]

Kottwitz, Jack J., Diane E. Preziosi, Margaret A. Miller, Jose A. Ramos-Vara, David J. Maggs, and John D. Bonagura. 2004. Heart Failure Caused by Toxoplasmosis in a Fennec Fox (Fennecus Zerda). *Journal of the American Animal Hospital Association* 40: 501–7. [CrossRef]

Kouam, Marc K., Anastasia Diakou, Vaia Kanzoura, Elias Papadopoulos, Alvin A. Gajadhar, and Georgios Theodoropoulos. 2010. A Seroepidemiological Study of Exposure to Toxoplasma, Leishmania, Echinococcus and Trichinella in Equids in Greece and Analysis of Risk Factors. *Veterinary Parasitology* 170: 170–75. [CrossRef] [PubMed]

Kovacova, D. 1993. Serologicka Prevelencia *Toxoplasma gondii* u Oviec. *Veterinarstvi* 43: 51–54.

Kramer, W. 1966. Frontiers of Neurological Diagnosis in Acquired Toxoplasmosis. *Psychiatria, Neurologia, Neurochirurgia* 69: 43–64. [PubMed]

Krasteva, D., Mylene Toubiana, Sri Hartati, Asmarani Kusumawati, Jean Francois Dubremetz, and Johannes Sri Widada. 2009. Development of Loop-Mediated Isothermal Amplification (LAMP) as a Diagnostic Tool of Toxoplasmosis. *Veterinary Parasitology* 162: 327–31. [CrossRef] [PubMed]

Kreuder, Christine, Melissa A. Miller, David A. Jessup, Linda J. Lowenstine, M. D. Harris, Jack A. Ames, Tim E. Carpenter, Patricia A. Conrad, and Jonna Ann Mazet. 2003. Patterns of Mortality in Southern Sea Otters (Enhydra Lutris Nereis) from 1998–2001. *Journal of Wildlife Diseases* 39: 495–509. [CrossRef]

Künzel, Frank, Andrea Gruber, Alexander Tichy, Renate Edelhofer, Barbara Nell, Jasmin Hassan, Michael Leschnik, Johann G. Thalhammer, and Anja Joachim. 2008. Clinical Symptoms and Diagnosis of Encephalitozoonosis in Pet Rabbits. *Veterinary Parasitology* 151: 115–24. [CrossRef] [PubMed]

Kuruca, Ljiljana, Ivana Klun, Aleksandra Uzelac, Aleksandra Nikolić, Branko Bobić, Stanislav Simin, Vesna Lalošević, Dušan Lalošević, and Olgica Djurković-Djaković. 2017. Detection of *Toxoplasma gondii* in Naturally Infected Domestic Pigs in Northern Serbia. *Parasitology Research* 116: 3117–23. [CrossRef]

Kutz, Susan J., Brett Elkin, Anne Gunn, and Jitender P. Dubey. 2000. Prevalence of *Toxoplasma gondii* Antibodies in Muskox (Ovibos Moschatus) Sera from Northern Canada. *The Journal of Parasitology* 86: 879–82. [CrossRef] [PubMed]

Lachaud, Laurence, Olivier Calas, Marie Christine Picot, Sahar Albaba, Nathalie Bourgeois, and Francine Pratlong. 2009. Value of 2 IgG Avidity Commercial Tests Used Alone or in Association to Date Toxoplasmosis Contamination. *Diagnostic Microbiology and Infectious Disease* 64: 267–74. [CrossRef]

Lachenmaier, Sabrina M., Mária A. Deli, Markus Meissner, and Oliver Liesenfeld. 2011. Intracellular Transport of *Toxoplasma gondii* through the Blood-Brain Barrier. *Journal of Neuroimmunology* 232: 119–30. [CrossRef] [PubMed]

Ladee, G. A. 1966. Diagnostic Problems in Psychiatry with Regard to Acquired Toxoplasmosis. *Psychiatria, Neurologia, Neurochirurgia* 69: 65–82.

Laforet, Celine Kaae, Gunita Deksne, Heidi Huus Petersen, Pikka Jokelainen, Maria Vang Johansen, and Brian Lassen. 2019. *Toxoplasma gondii* Seroprevalence in Extensively Farmed Wild Boars (Sus Scrofa) in Denmark. *Acta Veterinaria Scandinavica* 61: 4. [CrossRef] [PubMed]

Lago, Eleonor Gastal, Matteo Baldisserotto, Joao Rubiao Hoefel Filho, D. Santiago, and R. Jungblut. 2007. Agreement between Ultrasonography and Computed Tomography in Detecting Intracranial Calcifications in Congenital Toxoplasmosis. *Clinical Radiology* 62: 1004–11. [CrossRef]

Langoni, Helio, Haroldo Greca, Felipe F. Guimarães, Leila S. Ullmann, Fernanda C. Gaio, Robson S. Uehara, Eric P. Rosa, Rogério M. Amorim, and Rodrigo C. Da Silva. 2011. Serological Profile of *Toxoplasma gondii* and Neospora Caninum Infection in Commercial Sheep from São Paulo State, Brazil. *Veterinary Parasitology* 177: 50–54. [CrossRef]

Lass, Anna, Halina Pietkiewicz, E. Modzelewska, Aurelien Dumètre, Beata Szostakowska, and Przemyslaw Myjak. 2009. Detection of *Toxoplasma gondii* Oocysts in Environmental Soil Samples Using Molecular Methods. *European Journal of Clinical Microbiology & Infectious Diseases: Official Publication of the European Society of Clinical Microbiology* 28: 599–605. [CrossRef]

Lazăr, L., and I. Barbu. 2007. Suferința Feto-Placentară Și Dimorfismul Sechelar Neuro-Ocular În Toxoplasmoza Congenitală. *Revista Română de Parazitologie* 17: 106–7.

Leblebicier, Abdullah, and Kader Yıldız. 2014. [Seroprevalence of Toxoplasma gondii in sheep in Silopi district by using indirect fluorescent antibody test (IFAT)]. *Turkiye Parazitolojii Dergisi* 38: 1–4. [CrossRef] [PubMed]

Lee, Jung-Youn, Sang-Eun Lee, Eunggoo G. Lee, and Kunho Song. 2008. Nested PCR-Based Detection of *Toxoplasma gondii* in German Shepherd Dogs and Stray Cats in South Korea. *Research in Veterinary Science* 85: 125–27. [CrossRef] [PubMed]

Lee, Seung-Hun, Sang-Eun Lee, Min-Goo Seo, Youn-Kyoung Goo, Kwang-Hyun Cho, Gil-Jae Cho, Oh-Deog Kwon, Dongmi Kwak, and Won-Ja Lee. 2014. Evidence of *Toxoplasma gondii* Exposure among Horses in Korea. *The Journal of Veterinary Medical Science* 76: 1663–65. [CrossRef]

Li, Jian, Xiuzhen Yang, Dongchun Liang, and Peimei Liu. 2006. Study of a PCR-ELISA Assay for the Detection of *Toxoplasma gondii* DNA. *Chinese Journal Zoon* 22: 356–59.

Li, Yong-Nian, XinWen Nie, Qun-Yi Peng, Xiao-Qiong Mu, Ming Zhang, Meng-Yuan Tian, and Shao-ju Min. 2015. Seroprevalence and Genotype of *Toxoplasma gondii* in Pigs, Dogs and Cats from Guizhou Province, Southwest China. *Parasites & Vectors* 8: 214. [CrossRef]

Liang, Zhen-Bo, Yue-Ju Feng, Kai Li, Hui-Xi Lu, and Qiu-Feng Shen. 2006. Serological Survey on Toxoplasmosis in Huadu District of Guangzhou City in 2004. *Journal of Tropical Medicine* 6: 448–49.

Liassides, Marios, Vasiliki Christodoulou, Joanna Moschandreas, Christos Karagiannis, George Mitis, Maria Koliou, and Maria Antoniou. 2016. Toxoplasmosis in Female High School Students, Pregnant Women and Ruminants in Cyprus. *Transactions of the Royal Society of Tropical Medicine and Hygiene* 110: 359–66. [CrossRef]

Lima, Vanessa Yuri de, Helio Langoni, Aristeu Vieira da Silva, Sandia Bergamaschi Pezerico, André Peres Barbosa de Castro, Rodrigo Costa da Silva, and João Pessoa Araújo. 2011. Chlamydophila Psittaci and *Toxoplasma gondii* Infection in Pigeons (Columba Livia) from São Paulo State, Brazil. *Veterinary Parasitology* 175: 9–14. [CrossRef]

Lindová, Jitka, Martina Novotná, Jan Havlícek, Eva Jozífková, Anna Skallová, Petra Kolbeková, Zdenek Hodný, Petr Kodym, and Jaroslav Flegr. 2006. Gender Differences in Behavioural Changes Induced by Latent Toxoplasmosis. *International Journal for Parasitology* 36: 1485–92. [CrossRef] [PubMed]

Lindsay, David S., E. Jane Kelly, Richard D. McKown, Franklin J. Stein, John Plozer, James Herman, Byron L. Blagburn, and Jitender P. Dubey. 1996. Prevalence of Neospora Caninum and *Toxoplasma gondii* Antibodies in Coyotes (Canis Latrans) and Experimental Infections of Coyotes with Neospora Caninum. *The Journal of Parasitology* 82: 657–59. [CrossRef]

Lindsay, David S., and Louis M. Weiss, eds. 2004. *Opportunistic Infections: Toxoplasma, Sarcocystis, and Microsporidia: 9*, 2004th ed. Boston: Springer.

Lindsay, David S., Richard D. McKown, Jennifer A. DiCristina, Carly N. Jordan, Sheila Mitchell, David W. Oates, and Mauritz C. Sterner. 2005. Prevalence of Agglutinating Antibodies to *Toxoplasma gondii* in Adult and Fetal Mule Deer (Odocoileus Hemionus) from Nebraska. *The Journal of Parasitology* 91: 1490–91. [CrossRef] [PubMed]

Liu, Jingshuang, Jinzhong Cai, Wei Zhang, Qun Liu, Dan Chen, Jinpeng Han, and Q. R. Liu. 2008. Seroepidemiology of Neospora Caninum and *Toxoplasma gondii* Infection in Yaks (Bos Grunniens) in Qinghai, China. *Veterinary Parasitology* 152: 330–32. [CrossRef]

Lopes, Ana Patrícia, Ana Cristina Oliveira, Sara Granada, Filipa T. Rodrigues, Elias Papadopoulos, Henk Schallig, Jitender P. Dubey, and Luís Cardoso. 2017. Antibodies to *Toxoplasma gondii* and Leishmania Spp. in Domestic Cats from Luanda, Angola. *Veterinary Parasitology* 239: 15–18. [CrossRef] [PubMed]

Lopes, Ana Patrícia, Luís Cardoso, and Manuela Rodrigues. 2008. Serological Survey of *Toxoplasma gondii* Infection in Domestic Cats from Northeastern Portugal. *Veterinary Parasitology* 155: 184–89. [CrossRef] [PubMed]

Lopes, Ana Patrícia, Susana Sousa, J. P. Dubey, Ana J. Ribeiro, Ricardo Silvestre, Mário Cotovio, Henk D. F. H. Schallig, Luís Cardoso, and Anabela Cordeiro da Silva. 2013. Prevalence of Antibodies to Leishmania Infantum and *Toxoplasma gondii* in Horses from the North of Portugal. *Parasites & Vectors* 6: 178. [CrossRef]

Lopes, Welber Daniel Zanetti, Alvimar Jose da Costa, Fernando Augusto de Souza, Joao Domingos F. Rodrigues, Gustavo Henrique Nogueira Costa, Vando Edesio Soares, and Giane Serafim Silva. 2009. Semen Variables of Sheep (Ovis Aries) Experimentally Infected with *Toxoplasma gondii*. *Animal Reproduction Science* 111: 312–19. [CrossRef]

Lopes, Welber Daniel Zanetti, Thaís Rabelo dos Santos, Ricardo dos Santos da Silva, Walter Matheus Rossanese, Fernando Augusto de Souza, Joana D'Ark de Faria Rodrigues, Rafael Paranhos de Mendonça, Vando Edésio Soares, and Alvimar José da Costa. 2010. Seroprevalence of and Risk Factors for *Toxoplasma gondii* in Sheep Raised in the Jaboticabal Microregion, São Paulo State, Brazil. *Research in Veterinary Science* 88: 104–6. [CrossRef]

Lou, Zhi-Long, Wei Cong, Qing-Feng Meng, Wu-Wen Sun, Si-Yuan Qin, Xiao-Xuan Zhang, Xing-Quan Zhu, and Ai-Dong Qian. 2015. First Report of *Toxoplasma gondii* Seroprevalence in Farmed Arctic Foxes (Vulpes Lagopus) in Eastern and Northeastern China. *Vector Borne and Zoonotic Diseases (Larchmont, N.Y.)* 15: 775–78. [CrossRef] [PubMed]

Machacova, Tereza, Eva Bartova, Kamil Sedlak, Radka Slezakova, Marie Budikova, Diego Piantedosi, and Vincenzo Veneziano. 2016. Seroprevalence and Risk Factors of Infections with Neospora Caninum and *Toxoplasma gondii* in Hunting Dogs from Campania Region, Southern Italy. *Folia Parasitologica* 63: 12. [CrossRef] [PubMed]

Machado, Dália Monique Ribeiro, Luiz Daniel de Barros, Beatriz de Souza Lima Nino, Andressa de Souza Pollo, Ana Clécia Dos Santos Silva, Lívia Perles, Marcos Rogério André, Rosângela Zacarias Machado, João Luis Garcia, and Estevam Guilherme Lux Hoppe. 2021. *Toxoplasma gondii* Infection in Wild Boars (Sus Scrofa) from the State of São Paulo, Brazil: Serology, Molecular Characterization, and Hunter's Perception on Toxoplasmosis. *Veterinary Parasitology, Regional Studies and Reports* 23: 100534. [CrossRef] [PubMed]

MacLaren, A., Marcia Attias, and Wanderley de Souza. 2004. Aspects of the Early Moments of Interaction between Tachyzoites of *Toxoplasma gondii* with Neutrophils. *Veterinary Parasitology* 125: 301–12. [CrossRef] [PubMed]

Macpherson, Calum N. L. 2005. Human Behaviour and the Epidemiology of Parasitic Zoonoses. *International Journal for Parasitology* 35: 1319–31. [CrossRef]

Magalhães, Fernando J. R., Müller Ribeiro-Andrade, Fátima M. Souza, Carlos D. F. Lima Filho, Alexander Welker Biondo, Odilon Vidotto, Italmar Teodorico Navarro, and Rinaldo A. Mota. 2017. Seroprevalence and Spatial Distribution of *Toxoplasma gondii* Infection in Cats, Dogs, Pigs and Equines of the Fernando de Noronha Island, Brazil. *Parasitology International* 66: 43–46. [CrossRef]

Mainardi, Rodrigo Soares, José Rafael Modolo, Anee Valéria Mendonça Stachissini, Carlos Roberto Padovani, and Hélio Langoni. 2003. [Seroprevalence of Toxoplasma gondii in dairy goats in the São Paulo State, Brazil]. *Revista Da Sociedade Brasileira De Medicina Tropical* 36: 759–61. [CrossRef] [PubMed]

Mainar-Jaime, Raul C., and M. Barberán. 2007. Evaluation of the Diagnostic Accuracy of the Modified Agglutination Test (MAT) and an Indirect ELISA for the Detection of Serum Antibodies against *Toxoplasma gondii* in Sheep through Bayesian Approaches. *Veterinary Parasitology* 148: 122–29. [CrossRef] [PubMed]

Malmsten, Anna, Ulf Magnusson, Francisco Ruiz-Fons, David González-Barrio, and Anne-Marie Dalin. 2018. A Serologic Survey of Pathogens in Wild Boar (Sus Scrofa) In Sweden. *Journal of Wildlife Diseases* 54: 229–37. [CrossRef]

Malmsten, Jonas, Eva-Britt Jakubek, and Camilla Björkman. 2011. Prevalence of Antibodies against *Toxoplasma gondii* and Neospora Caninum in Moose (Alces Alces) and Roe Deer (Capreolus Capreolus) in Sweden. *Veterinary Parasitology* 177: 275–80. [CrossRef] [PubMed]

Mancianti, Francesca, Simona Nardoni, Gaetano Ariti, Dario Parlanti, Giovanna Giuliani, and Roberto A. Papini. 2010. Cross-Sectional Survey of *Toxoplasma gondii* Infection in Colony Cats from Urban Florence (Italy). *Journal of Feline Medicine and Surgery* 12: 351–54. [CrossRef] [PubMed]

Mancianti, Francesca, Simona Nardoni, Roberto Papini, Linda Mugnaini, Mina Martini, Iolanda Altomonte, Federica Salari, Carlo D'Ascenzi, and Jitender P. Dubey. 2014. Detection and Genotyping of *Toxoplasma gondii* DNA in the Blood and Milk of Naturally Infected Donkeys (Equus Asinus). *Parasites & Vectors* 7: 165. [CrossRef]

Marobin, Letícia, Maristela Lovato Flôres, Bárbara Bach Rizzatti, Stefanie Dickel Segabinazi, Vera Regina Albuquerque Lagaggio, Marcelo Grigulo, and Marcos Adriano Scalco. 2004. Prevalência de anticorpos para Toxoplasma gondii em emas (Rhea americana) em diferentes criatórios do Estado do Rio Grande do Sul. *Brazilian Journal of Veterinary Research and Animal Science* 41: 5–9. [CrossRef]

Marques, Luiz Carlos, Alvimar José da Costa, Carlos Wilson Gomes Lopes, Flávio Ruas de Moraes, and Julieta Rodine Engracia de Moraes. 1995. Toxoplasmose experimental em éguas gestantes: Estudo dos fetos e placentas. *Brazilian Journal of Veterinary Research and Animal Science* 32: 246–50. [CrossRef]

Martinez-Carasco, Carlos, J.M. Ortiz, A. Bernabe, M. Rocio Ruiz de Ybanez, Magdalena Garijo, and Francisco D. Alonso. 2004. Serological Respone of Red-Legged Partridges (Alectoris Rufa) after Oral Inoculation with *Toxoplasma gondii* Oocysts. *Veterinary Bulletin* 74: 1179.

Maruyama, Soichi, Shinya Hiraga, Eiji Yokoyama, Masayuki Naoi, Yuji Tsuruoka, Yoshihiro Ogura, Katsutoshi Tamura, Shinichi Namba, Yasuhiko Kameyama, Satoru Nakamura, and et al. 1998. Seroprevalence of Bartonella Henselae and *Toxoplasma gondii* Infections among Pet Cats in Kanagawa and Saitama Prefectures. *The Journal of Veterinary Medical Science* 60: 997–1000. [CrossRef]

Masala, Giovanna, Rosaura Porcu, Laura Madau, Antonio Tanda, B. Ibba, Giuseppe Satta, and Sebastiana Tola. 2003. Survey of Ovine and Caprine Toxoplasmosis by IFAT and PCR Assays in Sardinia, Italy. *Veterinary Parasitology* 117: 15–21. [CrossRef]

Masatani, Tatsunori, Yasuhiro Takashima, Masaki Takasu, Aya Matsuu, and Tomohiko Amaya. 2016. Prevalence of Anti-*Toxoplasma gondii* Antibody in Domestic Horses in Japan. *Parasitology International* 65: 146–50. [CrossRef]

Mason, Sam, Rupert J. Quinnell, and Judith E. Smith. 2010. Detection of *Toxoplasma gondii* in Lambs via PCR Screening and Serological Follow-Up. *Veterinary Parasitology* 169: 258–63. [CrossRef]

Massie, Gloeta N., Michael W. Ware, Eric N. Villegas, and Michael W. Black. 2010. Uptake and Transmission of *Toxoplasma gondii* Oocysts by Migratory, Filter-Feeding Fish. *Veterinary Parasitology* 169: 296–303. [CrossRef] [PubMed]

Matsuo, Kayoko, Rika Kamai, Hirona Uetsu, Hanyu Goto, Yasuhiro Takashima, and Kisaburo Nagamune. 2014. Seroprevalence of *Toxoplasma gondii* Infection in Cattle, Horses, Pigs and Chickens in Japan. *Parasitology International* 63: 638–39. [CrossRef] [PubMed]

Maubon, Danièle, Daniel Ajzenberg, Marie-Pierre Brenier-Pinchart, Marie-Laure Dardé, and Hervé Pelloux. 2008. What Are the Respective Host and Parasite Contributions to Toxoplasmosis? *Trends in Parasitology* 24: 299–303. [CrossRef]

McAllister, Milton M. 2005. A Decade of Discoveries in Veterinary Protozoology Changes Our Concept of "Subclinical" Toxoplasmosis. *Veterinary Parasitology* 132: 241–47. [CrossRef] [PubMed]

Measures, Lena N., Jitender P. Dubey, P. Labelle, and D. Martineau. 2004. Seroprevalence of *Toxoplasma gondii* in Canadian Pinnipeds. *Journal of Wildlife Diseases* 40: 294–300. [CrossRef]

Mecca, Juliana Nunes, Luciana Regina Meireles, and Heitor Franco de Andrade. 2011. Quality Control of *Toxoplasma gondii* in Meat Packages: Standardization of an ELISA Test and Its Use for Detection in Rabbit Meat Cuts. *Meat Science* 88: 584–89. [CrossRef] [PubMed]

Mederle, Narcisa, Gheorghe Dărăbuş, Ionela Hotea, and Ovidiu Mederle. 2007. Surse Şi Modalităţi de Infestare În Toxoplasmoză. *Revista Română de Parazitologie* 17: 119.

Meireles, Luciana Regina, Andres Jimenez Galisteo, Eduardo Pompeu, and Heitor Franco de Andrade Jr. 2004. *Toxoplasma gondii* Spreading in an Urban Area Evaluated by Seroprevalence in Free-Living Cats and Dogs. *Tropical Medicine & International Health: TM & IH* 9: 876–81. [CrossRef]

Meireles, Luciana Regina, Andrés Jimenez Galisteo Jr., and Heitor Franco de Andrade Jr. 2003. Serological Survey of Antibodies to *Toxoplasma gondii* in Food Animals from São Paulo State, Brazil. *Brazilian Journal of Veterinary Research and Animal Science* 40: 267–71. [CrossRef]

de Melo, Renata Pimental Bandeira, Jonatas Campos de Almeida, Debora Costa Viegas de Lima, Camila de Morais Pedrosa, Fernando Jorge Rodrigues Magalhães, Adrianne Mota Alcântara, Luiz Daniel de Barros, Rafael F. C. Vieira, Joao Luis Garcia, and Rinaldo Aparecido Mota. 2016. Atypical *Toxoplasma gondii* Genotype in Feral Cats from the Fernando de Noronha Island, Northeastern Brazil. *Veterinary Parasitology* 224: 92–95. [CrossRef] [PubMed]

Messier, Valerie, B. Lévesque, Jean Francois Proulx, Louis Rochette, Michael D. Libman, Brian J. Ward, Bouchra Serhir, Michel Couillard, Nicholas H. Ogden, Eric Dewailly, and et al. 2009. Seroprevalence of *Toxoplasma gondii* among Nunavik Inuit (Canada). *Zoonoses and Public Health* 56: 188–97. [CrossRef] [PubMed]

Michalski, Mirosław, and Platt-Aleksandra Samoraj. 2004. Extent of *Toxoplasma gondii* Invasion in Goat and Sheep from the Olsztyn Region. *Medycyna Weterynaryjna* 60: 70–71.

Michalski, Mirosław, and Aleksandra Platt-Samoraj. 2004. [Prevalence of Toxoplasma gondii infection in cats from Olsztyn city]. *Wiadomosci Parazytologiczne* 50: 303–5.

Millán, Javier, Oscar Cabezón, Marcela Pabón, Jitender P. Dubey, and Sonia Almería. 2009. Seroprevalence of *Toxoplasma gondii* and Neospora Caninum in Feral Cats (Felis Silvestris Catus) in Majorca, Balearic Islands, Spain. *Veterinary Parasitology* 165: 323–26. [CrossRef]

Miller, Melissa Ann, Michael E. Grigg, Christine Kreuder, Erick R. James, Ann C. Melli, Paul R. Crosbie, David A. Jessup, John C. Boothroyd, Deborah Brownstein, and Patricia A. Conrad. 2004. An Unusual Genotype of *Toxoplasma gondii* Is Common in California Sea Otters (Enhydra Lutris Nereis) and Is a Cause of Mortality. *International Journal for Parasitology* 34: 275–84. [CrossRef] [PubMed]

Miller, Melissa Ann, W A. Miller, Patricia A. Conrad, Erick R. James, Ann C. Melli, Christian M. Leutenegger, Haydee A. Dabritz, Andrea E. Packham, David M. Paradies, Michael D. Harris, and et al. 2008a. Type X *Toxoplasma gondii* in a Wild Mussel and Terrestrial Carnivores from Coastal California: New Linkages between Terrestrial Mammals, Runoff and Toxoplasmosis of Sea Otters. *International Journal for Parasitology* 38: 1319–28. [CrossRef] [PubMed]

Miller, Melissa, Patricia Conrad, E. R. James, Andrea Packham, Sharon Toy-Choutka, Michael J. Murray, David Jessup, and Michael Grigg. 2008b. Transplacental Toxoplasmosis in a Wild Southern Sea Otter (Enhydra Lutris Nereis). *Veterinary Parasitology* 153: 12–18. [CrossRef] [PubMed]

Mineo, Tiago W. P., Deise Aparecida Oliveira Silva, Katarina Näslund, Camilla Björkman, Arvid Uggla, and Jose Roberto Mineo. 2004. *Toxoplasma gondii* and Neospora Caninum Serological Status of Different Canine Populations from Uberlândia, Minas Gerais. *Arquivo Brasileiro de Medicina Veterinária e Zootecnia* 56: 414–17. [CrossRef]

Mircean, Viorica, Adriana Titilincu, and Cozma Vasile. 2010. Prevalence of Endoparasites in Household Cat (Felis Catus) Populations from Transylvania (Romania) and Association with Risk Factors. *Veterinary Parasitology* 171: 163–66. [CrossRef] [PubMed]

Miró, Guadalupe, Ana Montoya, Santos Jiménez, Carolina Frisuelos, Marta Mateo, and Isabel Fuentes. 2004. Prevalence of Antibodies to *Toxoplasma gondii* and Intestinal Parasites in Stray, Farm and Household Cats in Spain. *Veterinary Parasitology* 126: 249–55. [CrossRef]

Miró, Guadalupe, Cristina Rupérez, Rocío Checa, Rosa Gálvez, Leticia Hernández, Manuel García, Isabel Canorea, Valentina Marino, and Ana Montoya. 2014. Current Status of L. Infantum Infection in Stray Cats in the Madrid Region (Spain): Implications for the Recent Outbreak of Human Leishmaniosis? *Parasites & Vectors* 7: 112. [CrossRef]

Mitchell, M. A., Laura L. Hungeford, C. Nixon, T. Esker, J. Sullivan, Robert Koerkenmeier, and Jitender P. Dubey. 1999. Serologic Survey for Selected Infectious Disease Agents in Raccoons from Illinois. *Journal of Wildlife Diseases* 35: 347–55. [CrossRef]

Montoya, Ana, Manuel García, Rosa Gálvez, Rocio Checa, Valentina Marino, Juliana Sarquis, Juan Pedro Barrera, Cristina Ruperez, L. Caballero, Carmen Chicharro, and et al. 2018. Implications of Zoonotic and Vector-Borne Parasites to Free-Roaming Cats in Central Spain. *Veterinary Parasitology* 251: 125–30. [CrossRef]

Montoya, Ana, Guadalupe Miró, Marta Mateo, Carmen Ramírez, and Isabel Fuentes. 2009. Detection of *Toxoplasma gondii* in Cats by Comparing Bioassay in Mice and Polymerase Chain Reaction (PCR). *Veterinary Parasitology* 160: 159–62. [CrossRef]

Montoya, Jose G. 2002. Laboratory Diagnosis of *Toxoplasma gondii* Infection and Toxoplasmosis. *The Journal of Infectious Diseases* 185: S73–S82. [CrossRef]

Moore, Dadin Prando, Javier Regidor-Cerrillo, Eleonora Morrell, M. A. Poso, Dora B. Cano, Maria Rosa Leunda, L. Linschinky, Anselmo Carlos Odeon, Ernesto R. Odriozola, Luis Miguel Ortega-Mora, and et al. 2008. The Role of Neospora Caninum and *Toxoplasma gondii* in Spontaneous Bovine Abortion in Argentina. *Veterinary Parasitology* 156: 163–67. [CrossRef]

Moré, Gaston, Lais Pardini, Walter Basso, Mariana Alejandra Machuca, Diana Bacigalupe, Maria Cecilia Villanueva, Gereon Schares, Maria Cecilia Venturini, and Lucila Venturini. 2010. Toxoplasmosis and Genotyping of *Toxoplasma gondii* in Macropus Rufus and Macropus Giganteus in Argentina. *Veterinary Parasitology* 169: 57–61. [CrossRef] [PubMed]

Moré, Gaston, Lais Pardini, Walter Basso, Raul Marín, Diana Bacigalupe, G. Auad, Lucila Venturini, and Maria Cecilia Venturini. 2008a. Seroprevalence of Neospora Caninum, *Toxoplasma gondii* and Sarcocystis Sp. in Llamas (Lama Glama) from Jujuy, Argentina. *Veterinary Parasitology* 155: 158–60. [CrossRef]

Moré, Gaston, Walter Basso, Diana Bacigalupe, Maria Cecilia Venturini, and Lucila Venturini. 2008b. Diagnosis of Sarcocystis Cruzi, Neospora Caninum, and *Toxoplasma gondii* Infections in Cattle. *Parasitology Research* 102: 671–75. [CrossRef]

Morshedi, A., A. Mahmoodian, and A. Awaz. 2006. Determination of the Lowest Valuable Titre (Cut-off Titre) of Antitoxoplasma Antibodies for Detection of Toxoplasmosis in Sheep by ELISA and IFA Tests. *Veterinary Bulletin* 76: 1388.

Moskwa, Bozena, Aleksandra Kornacka, Aleksandra Cybulska, Wladyslaw Cabaj, Katarina Reiterova, Marek Bogdaszewski, Zaneta Steiner-Bogdaszewska, and Justyna Bien. 2018. Seroprevalence of *Toxoplasma gondii* and Neospora Caninum Infection in Sheep, Goats, and Fallow Deer Farmed on the Same Area. *Journal of Animal Science* 96: 2468–73. [CrossRef]

Mucker, Eric M., Jitender P. Dubey, Matthew J. Lovallo, and Jan G. Humphreys. 2006. Seroprevalence of Antibodies to *Toxoplasma gondii* in the Pennsylvania Bobcat (Lynx Rufus Rufus). *Journal of Wildlife Diseases* 42: 188–91. [CrossRef] [PubMed]

Munhoz, Alexandre Dias, Monia Andrade Souza, Sonia Carmen Lopo Costa, Jéssica de Souza Freitas, Aísla Nascimento da Silva, Luciana Carvalho Lacerda, Rebeca Dálety Santos Cruz, George Rêgo Albuquerque, and Maria Julia Salim Pereira. 2019. Factors Associated with the Distribution of Natural *Toxoplasma gondii* Infection among Equids in Northeastern Brazil. *Revista Brasileira De Parasitologia Veterinaria = Brazilian Journal of Veterinary Parasitology: Orgao Oficial Do Colegio Brasileiro De Parasitologia Veterinaria* 28: 283–90. [CrossRef] [PubMed]

Muñoz-Zanzi, Claudia, R. Tamayo, J. Balboa, and D. Hill. 2012. Detection of Oocyst-Associated Toxoplasmosis in Swine from Southern Chile. *Zoonoses and Public Health* 59: 389–92. [CrossRef]

Murata, K., K. Mizuta, K. Imazu, F. Terasawa, M. Taki, and T. Endoh. 2004. The Prevalence of *Toxoplasma gondii* Antibodies in Wild and Captive Cetaceans from Japan. *The Journal of Parasitology* 90: 896–98. [CrossRef]

Murphy, Thomas M., Julia Walochnik, Andreas Hassl, John Patrick Moriarty, Jean Mooney, Donal P. Toolan, Cosme Sanchez-Miguel, Aonghus J. O'Loughlin, and A. McAuliffe. 2007. Study on the Prevalence of *Toxoplasma gondii* and Neospora Caninum and Molecular Evidence of Encephalitozoon Cuniculi and Encephalitozoon (Septata) Intestinalis Infections in Red Foxes (Vulpes Vulpes) in Rural Ireland. *Veterinary Parasitology* 146: 227–34. [CrossRef]

Must, Kärt, Marjo K. Hytönen, Toomas Orro, Hannes Lohi, and Pikka Jokelainen. 2017. *Toxoplasma gondii* Seroprevalence Varies by Cat Breed. *PLoS ONE* 12: e0184659. [CrossRef] [PubMed]

Muz, Mustafa Necati, N. Altug, and Muhammet Karakavuk. 2013. Seroprevalence of *T. gondii* in Dairy Ruminant Production Systems, Shepherd Dogs among the Herds and Detection of *T. gondii*-like Oocyst in Cat Feces in Hatay Region. *AVKAE Dergisi* 3: 38–45.

Nagy, Bálint, Zoltán Bán, Artúr Beke, Gyula Richard Nagy, Levente Lázár, Csaba Papp, Erno Tóth-Pál, and Zoltán Papp. 2006. Detection of *Toxoplasma gondii* from Amniotic Fluid, a Comparison of Four Different Molecular Biological Methods. *Clinica Chimica Acta; International Journal of Clinical Chemistry* 368: 131–37. [CrossRef] [PubMed]

Navarrete, Maylín González, Matheus Dias Cordeiro, Yasmín Batista, Julio Cesar Alonso, Mário Márquez, Eugênio Roque, and Adivaldo Fonseca. 2017. Serological Detection of *Toxoplasma gondii* in Domestic Dogs in the Western Region of Cuba. *Veterinary Parasitology, Regional Studies and Reports* 9: 9–12. [CrossRef] [PubMed]

Naves, Christiana Savastano, Fernando Antonio Ferreira, Francisco de Sales Resende Carvalho, and Gustavo Henrique Nogueira Costa. 2005. Soroprevalência Da Toxoplasmose Em Equinos da Raça Mangalarga Marchador No Município De Uberlândia, Minas Gerais. *Veterinária Notícias* 11: 1.

Neagoe, Ionela Mirela, R. Gherman, A. Rădulescu, M. Pop, N. Bucurenci, A. Dumitrache, and Dan Steriu. 2007. Stabilirea Unei Posibile Transmiteri Congenitale a Toxoplasmozei Prin Depistarea Antigenelor Specifice (NTP-Hidrolaza) Cu Anticorpi Monoclonali La Gravidele Cu Seroconversie. *Revista Română de Parazitologie* 17: 43–48.

Negash, Tu'Umay, Tilahun Getachew, S. Patton, F. Prévot, and Philippe Dorchies. 2004. Serological Survey on Toxoplasmosis in Sheep and Goats in Nazareth, Ethiopia. *Revue de Medecine Veterinaire* 155: 486–87.

Nematollahi, Ahmad, Parisa Shahbazi, and Sara Nosrati. 2014. Serological Survey of Antibodies to *Toxoplasma gondii* in Sheep in North-West of Iran. *International Journal of Advanced Biological and Biomedical Research* 2: 2605–8.

Nerad, Thomas A., and Pierre-Marc Daggett. 1992. Cultivation of Toxoplasma. In *Protocols in Protozoology*. A-38.1-A-38.3. Lawrence: Society of Protozoologists.

Nesbakken, Truls 2009. Food Safety in a Global Market—Do We Need to Worry? *Small Ruminant Research* 86: 63–66. [CrossRef]

Neto, José O. Araújo, Sérgio S. Azevedo, Solange M. Gennari, Mikaela R. Funada, Hilda F. J. Pena, Ana Raquel C. P. Araújo, and Carolina S. A. Batista. 2008. Prevalence and Risk Factors for Anti-*Toxoplasma gondii* Antibodies in Goats of the Seridó Oriental Microregion, Rio Grande Do Norte State, Northeast Region of Brazil. *Veterinary Parasitology* 156: 329–32. [CrossRef] [PubMed]

Neves, Michelle, Ana Patrícia Lopes, Carolina Martins, Raquel Fino, Cláudia Paixão, Liliana Damil, and Clara Lima. 2020. Survey of Dirofilaria Immitis Antigen and Antibodies to Leishmania Infantum and *Toxoplasma gondii* in Cats from Madeira Island, Portugal. *Parasites & Vectors* 13: 117. [CrossRef]

Nguyen, Thuy Thi-Dieu, Se-Eun Choe, Jae-Won Byun, Hong-Bum Koh, Hee-Soo Lee, and Seung-Won Kang. 2012. Seroprevalence of *Toxoplasma gondii* and Neospora Caninum in Dogs from Korea. *Acta Parasitologica* 57: 7–12. [CrossRef] [PubMed]

Nogami, Sadao, Akihisa Tabata, Tadaaki Moritomo, and Y. Hayashi. 1999. Prevalence of Anti-*Toxoplasma gondii* Antibody in Wild Boar, Sus Scrofa Riukiuanus, on Iriomote Island, Japan. *Veterinary Research Communications* 23: 211–14. [CrossRef] [PubMed]

Nutter, Felicia B., Jitender P. Dubey, Jay F. Levine, Edward B. Breitschwerdt, Richard B. Ford, and Michael K. Stoskopf. 2004. Seroprevalences of Antibodies against Bartonella Henselae and *Toxoplasma gondii* and Fecal Shedding of *Cryptosporidium* Spp., *Giardia* Spp., and *Toxocara cati* in Feral and Pet Domestic Cats. *Journal of the American Veterinary Medical Association* 225: 1394–98. [CrossRef]

Ogawa, Liza, Italmar Teodorico Navarro, Roberta Lemos Freire, R.C. DeOliveira, and Odilon Vidotto. 2005. Occurrence of Antibodies to *Toxoplasma gondii* in Sheep from the Londrina Region of the Paraná State, Brazil. *Veterinary Bulletin* 75: 78.

Olariu, Rares Teodor, O. Cretu, Gh Dărăbuş, I. Marincu, O. Jurovits, V. Erdelean, and L. Tirnea. 2008a. Screening for *Toxoplasma gondii* Antibodies among Women of Childbearing Age, in Timis County, Romania. *Abstracts of the 13th International Congres on Infectious Diseases. Malaysia* 13: 115.

Olariu, Rares Teodor. 2003. *Textbook of Medical Parasitology*. Timişoara: Eurostampa Publishing House.

Olariu, Teodor Rares, Octavian Creţu, Andrea Koreck, and Cristina Petrescu. 2006. Diagnosis of Toxoplasmosis in Pregnancy: Importance of Immunoglobulin G Avidity Test. *Roumanian Archives of Microbiology and Immunology* 65: 131–34. [PubMed]

Olariu, Teodor, Gheorghe Darabus, Octavian Cretu, Oana Jurovits, Eugenia Giura, Vlad Erdelean, Iosif Marincu, Ioan Iacobiciu, Cristina Petrescu, and Andrea Koreck. 2008b. Prevalence of *Toxoplasma gondii* Antibodies among Women of Childbearing Age in Timis County. *Lucrări Stiinţifice Medicină Veterinară* 41: 367–71.

Omata, Yoshitaka, Kouichi Murata, Kunio Ito, and And Naotaka Ishiguro. 2005a. Antibodies to *Toxoplasma gondii* in Free-Ranging Wild Boar (Sus Scrofa Leucomystax) in Shikoku, Japan. *Japanese Journal of Zoo and Wildlife Medicine* 10: 99–102. [CrossRef]

Omata, Yoshitaka, Naotaka Ishiguro, Rika Kano, Yuko Masukata, Akimasa Kudo, Hikaru Kamiya, and Haruka Fukui. 2005b. Prevalence of *Toxoplasma gondii* and Neospora Caninum in Sika Deer from Eastern Hokkaido, Japan. *Journal of Wildlife Diseases* 41: 454–58. [CrossRef] [PubMed]

Oncel, Taraneh, and Gulay Vural. 2006. Occurrence of *Toxoplasma gondii* Antibodies in Sheep in Istanbul, Turkey. *Veterinarski Arhiv* 76: 547–53.

Oncel, Taraneh, Gülay Vural, Cahit Babür, and Selçuk Kiliç. 2005. Detection of Toxoplasmosis Gondii Seropositivity in Sheep in Yalova by Sabin Feldman Dye Test and Latex Agglutination Test. *Turkiye Parazitolojii Dergisi* 29: 10–12.

Onyiche, ThankGod Emmanuel, and Isaiah Oluwafemi Ademola. 2015. Seroprevalence of Anti-*Toxoplasma gondii* Antibodies in Cattle and Pigs in Ibadan, Nigeria. *Journal of Parasitic Diseases: Official Organ of the Indian Society for Parasitology* 39: 309–14. [CrossRef] [PubMed]

Opsteegh, Marieke, Peter Teunis, Lothar Züchner, Ad Koets, Merel Langelaar, and Joke van der Giessen. 2011. Low Predictive Value of Seroprevalence of *Toxoplasma gondii* in Cattle for Detection of Parasite DNA. *International Journal for Parasitology* 41: 343–54. [CrossRef]

Opsteegh, Marieke, Peter Teunis, Marieke Mensink, Lothar Züchner, Adriana Titilincu, Merel Langelaar, and Joke van der Giessen. 2010. Evaluation of ELISA Test Characteristics and Estimation of *Toxoplasma gondii* Seroprevalence in Dutch Sheep Using Mixture Models. *Preventive Veterinary Medicine* 96: 232–40. [CrossRef]

Ortega, Ynes R., Geraldine Saavedra, and Maria P. Torres. 2003. *Seroprevalence of Toxoplasma and Neospora in Cattle and Swine*. Griffin: Center for Food Safety, University of Georgia, UGA.

Ouslimani, Sabrine Fazia, Safia Tennah, Naouelle Azzag, Salima Yamina Derdour, Bernard China, and Farida Ghalmi. 2019. Seroepidemiological Study of the Exposure to *Toxoplasma gondii* among Horses in Algeria and Analysis of Risk Factors. *Veterinary World* 12: 2007–16. [CrossRef]

Pablos-Tanarro, Alba, Luis Miguel Ortega-Mora, Antonio Palomo, Francisco Casasola, and Ignacio Ferre. 2018. Seroprevalence of *Toxoplasma gondii* in Iberian Pig Sows. *Parasitology Research* 117: 1419–24. [CrossRef] [PubMed]

Pagmadulam, Baldorj, Punsantsogvoo Myagmarsuren, Naoaki Yokoyama, Badgar Battsetseg, and Yoshifumi Nishikawa. 2020. Seroepidemiological Study of Toxoplama Gondii in Small Ruminants (Sheep and Goat) in Different Provinces of Mongolia. *Parasitology International* 74: 101996. [CrossRef]

Palm, Christian, Hayrettin Tumani, Tim Pietzcker, and Dietmar Bengel. 2008. Diagnosis of Cerebral Toxoplasmosis by Detection of *Toxoplasma gondii* Tachyzoites in Cerebrospinal Fluid. *Journal of Neurology* 255: 939–41. [CrossRef] [PubMed]

Panadero, Rosario, A. Painceira, Ceferino López, Luis Vázquez, Adolfo Paz, Pablo Díaz, Vicente Dacal, S. Cienfuegos, Gonzalo Fernandez, Noelia Lago, and et al. 2010. Seroprevalence of *Toxoplasma gondii* and Neospora Caninum in Wild and Domestic Ruminants Sharing Pastures in Galicia (Northwest Spain). *Research in Veterinary Science* 88: 111–15. [CrossRef] [PubMed]

Papatsiros, Vasileios G., Labrini V. Athanasiou, Despina Stougiou, Elias Papadopoulos, Giorgios G. Maragkakis, Panagiotis D. Katsoulos, and Menelaos Lefkaditis. 2016. Cross-Sectional Serosurvey and Risk Factors Associated with the Presence of *Toxoplasma gondii* Antibodies in Pigs in Greece. *Vector Borne and Zoonotic Diseases (Larchmont, N.Y.)* 16: 48–53. [CrossRef] [PubMed]

Papini, Roberto Amerigo, Gloria Buzzone, Simona Nardoni, Guido Rocchigiani, and Francesca Mancianti. 2015. Seroprevalence and Genotyping of *Toxoplasma gondii* in Horses Slaughtered for Human Consumption in Italy. *Journal of Equine Veterinary Science* 35: 657–61. [CrossRef]

Parameswaran, N., Richard Christopher Andrew Thompson, Nandu Sundar, S. Pan, M. Johnson, Nicholas C. Smith, and Michael E. Grigg. 2010. Non-Archetypal Type II-like and Atypical Strains of *Toxoplasma gondii* Infecting Marsupials of Australia. *International Journal for Parasitology* 40: 635–40. [CrossRef] [PubMed]

Parameswaran, N., Ryan M. O'Handley, Michael E. Grigg, Stanley Gordon Fenwick, and Richard Christopher Andrew Thompson. 2009. Seroprevalence of *Toxoplasma gondii* in Wild Kangaroos Using an ELISA. *Parasitology International* 58: 161–65. [CrossRef] [PubMed]

Pardini, Lais, Gastón Moré, Marcelo Rudzinski, María L. Gos, Lucía M. Campero, Alejandro Meyer, Mariana Bernstein, Juan M. Unzaga, and María C. Venturini. 2016. *Toxoplasma gondii* Isolates from Chickens in an Area with Human Toxoplasmic Retinochoroiditis. *Experimental Parasitology* 166: 16–20. [CrossRef]

Paştiu, Anamaria Ioana, Daniel Ajzenberg, Adriana Györke, Ovidiu Şuteu, Anamaria Balea, Benjamin M. Rosenthal, Zsuzsa Kalmár, Cristian Domşa, and Vasile Cozma. 2015a. Traditional Goat Husbandry May Substantially Contribute to Human Toxoplasmosis Exposure. *The Journal of Parasitology* 101: 45–49. [CrossRef] [PubMed]

Paştiu, Anamaria Ioana, Adriana Györke, Radu Blaga, Viorica Mircean, Benjamin Martin Rosenthal, and Vasile Cozma. 2013. In Romania, Exposure to *Toxoplasma gondii* Occurs Twice as Often in Swine Raised for Familial Consumption as in Hunted Wild Boar, but Occurs Rarely, If Ever, among Fattening Pigs Raised in Confinement. *Parasitology Research* 112: 2403–7. [CrossRef]

Paştiu, Anamaria Ioana, Adriana Györke, Zsuzsa Kalmár, Pompei Bolfă, Benjamin Martin Rosenthal, Miruna Oltean, Isabelle Villena, Marina Spînu, and Vasile Cozma. 2015b. *Toxoplasma gondii* in Horse Meat Intended for Human Consumption in Romania. *Veterinary Parasitology* 212: 393–95. [CrossRef]

Paştiu, Anamaria Ioana, Anamaria Cozma-Petruţ, Aurélien Mercier, Anamaria Balea, Lokman Galal, Viorica Mircean, Dana Liana Pusta, Liviu Bogdan, and Adriana Györke. 2019. Prevalence and Genetic Characterization of *Toxoplasma gondii* in Naturally Infected Backyard Pigs Intended for Familial Consumption in Romania. *Parasites & Vectors* 12: 586. [CrossRef]

Pereira-Bueno, Juana Maria, Agueda Quintanilla-Gozalo, Valentin Pérez-Pérez, Gema Alvarez-García, Esther Collantes-Fernández, and Luis Miguel Ortega-Mora. 2004. Evaluation of Ovine Abortion Associated with *Toxoplasma gondii* in Spain by Different Diagnostic Techniques. *Veterinary Parasitology* 121: 33–43. [CrossRef]

Petersen, Eskild, Arnold Pollak, and Ingrid Reiter-Owona. 2001. Recent Trends in Research on Congenital Toxoplasmosis. *International Journal for Parasitology* 31: 115–44. [CrossRef]

Petersen, Eskild, Benjamin Edvinsson, B. Lundgren, Thomas Benfield, and Birgitta Evengård. 2006. Diagnosis of Pulmonary Infection with *Toxoplasma gondii* in Immunocompromised HIV-Positive Patients by Real-Time PCR. *European Journal of Clinical Microbiology & Infectious Diseases: Official Publication of the European Society of Clinical Microbiology* 25: 401–4. [CrossRef]

Petithory, Jean Claude, Ingrid Reiter-Owona, F. Berthelot, M. Milgram, J. De Loye, and E. Petersen. 1996. Performance of European Laboratories Testing Serum Samples for *Toxoplasma gondii*. *European Journal of Clinical Microbiology & Infectious Diseases: Official Publication of the European Society of Clinical Microbiology* 15: 45–49. [CrossRef]

Piasecki, Tomasz, E. Śmielewska-Łoś, and Alina Wieliczko. 2004. Prevalence of Antibodies to *Toxoplasma gondii* in Urban Pigeons and Goshawks. *Medycyna Weterynaryjna* 60: 72–75.

Pietkiewicz, Halina, Elzbieta Hiszczyńska-Sawicka, Jozef Kur, E. Petersen, Henrik Vedel Nielsen, M. Paul, Miroslaw Stankiewicz, and Przemysllaw Myjak. 2007. Usefulness of *Toxoplasma gondii* Recombinant Antigens (GRA1, GRA7 and SAG1) in an Immunoglobulin G Avidity Test for the Serodiagnosis of Toxoplasmosis. *Parasitology Research* 100: 333–37. [CrossRef]

Pinon, Jean Michel, Henri Dumon, Cathy Chemla, Jacqueline Franck, Eskild Petersen, Morten Lebech, Jade Zufferey, Marie Helene Bessieres, Pierre Marty, R. Holliman, and et al. 2001. Strategy for Diagnosis of Congenital Toxoplasmosis: Evaluation of Methods Comparing Mothers and Newborns and Standard Methods for Postnatal Detection of Immunoglobulin G, M, and A Antibodies. *Journal of Clinical Microbiology* 39: 2267–71. [CrossRef]

Pipia, A. Paola, Antonio Varcasia, Giorgia Dessì, Romina Panzalis, C. Gai, Francesca Nonnis, Fabrizia Veronesi, Claudia Tamponi, and Antonio Scala. 2018. Seroepidemiological and Biomolecular Survey on *Toxoplasma gondii* Infection on Organic Pig Farms. *Parasitology Research* 117: 1637–41. [CrossRef] [PubMed]

Porqueddu, M., Antonio Scala, and V. Tilocca. 2004. Principal Endoparasitoses of Domestic Cats in Sardinia. *Veterinary Research Communications* 28: 311–13. [CrossRef]

Powell, L. F., Tanya E. A. Cheney, Susanna Williamson, Edward Guy, Richard P. Smith, and Robert H. Davies. 2016. A Prevalence Study of Salmonella Spp., Yersinia Spp., *Toxoplasma gondii* and Porcine Reproductive and Respiratory Syndrome Virus in UK Pigs at Slaughter. *Epidemiology and Infection* 144: 1538–49. [CrossRef] [PubMed]

Prestrud, Kristin Wear, Kjetil Asbakk, Eva Fuglei, Torill Mørk, Audun Stien, Erik Ropstad, Morten Tryland, Geir Wing Gabrielsen, Christian Lydersen, Kit M. Kovacs, and et al. 2007. Serosurvey for *Toxoplasma gondii* in Arctic Foxes and Possible Sources of Infection in the High Arctic of Svalbard. *Veterinary Parasitology* 150: 6–12. [CrossRef] [PubMed]

Prestrud, Kristin Wear, Kjetil Asbakk, Torill Mørk, Eva Fuglei, Morten Tryland, and Chunlei Su. 2008. Direct High-Resolution Genotyping of *Toxoplasma gondii* in Arctic Foxes (Vulpes Lagopus) in the Remote Arctic Svalbard Archipelago Reveals Widespread Clonal Type II Lineage. *Veterinary Parasitology* 158: 121–28. [CrossRef] [PubMed]

Putignani, Lorenza, Livia Mancinelli, Federica Del Chierico, Donato Menichella, Daniel Adlerstein, Maria Cristina Angelici, Marianna Marangi, Federica Berrilli, Monica Caffara, D. Antonio Frangipane di Regalbono, and et al. 2011. Investigation of *Toxoplasma gondii* Presence in Farmed Shellfish by Nested-PCR and Real-Time PCR Fluorescent Amplicon Generation Assay (FLAG). *Experimental Parasitology* 127: 409–17. [CrossRef]

Qian, Weifeng, Hui Wang, Chunlei Su, Dan Shan, Xia Cui, Na Yang, Chaochao Lv, and Qun Liu. 2012. Isolation and Characterization of *Toxoplasma gondii* Strains from Stray Cats Revealed a Single Genotype in Beijing, China. *Veterinary Parasitology* 187: 408–13. [CrossRef] [PubMed]

Qiu, Hong-Yu, Xiao-Xuan Zhang, Jing Jiang, Yanan Cai, Peng Xu, Quan Zhao, and Chun-Ren Wang. 2020. *Toxoplasma gondii* Seropositivity and Associated Risk Factors in Cats (Felis Catus) in Three Provinces in Northeastern China from 2013 to 2019. *Vector Borne and Zoonotic Diseases (Larchmont, N.Y.)* 20: 723–27. [CrossRef] [PubMed]

Radbea, Narcisa, Ovidiu Mederle, D. Barbu, and D. Izvernariu. 2006a. Semnalarea Unui Caz de Corioretinită Cu *Toxoplasma gondii* La Om. *Lucr Stiint, Ser Med Vet Inst Agron Timisoara* 39: 17–21.

Radbea, Narcisa, Ovidiu Mederle, Gheorghe Dărăbuş, Ion Oprescu, Sorin Morariu, Marius Ilie, and Ionela Hotea. 2006b. Modificări Vasculare Ale Placentei Umane Infestată Cu *Toxoplasma gondii*. *Lucr Stiint, Ser Med Vet Inst Agron Timisoara* 39: 22–25.

Ragozo, Alessandra Mara Alves, Lucia Eiko Oishi Yai, L. N. Oliveira, Ricardo Augusto Dias, Jitender P. Dubey, and Solange Maria Gennari. 2008. Seroprevalence and Isolation of *Toxoplasma gondii* from Sheep from São Paulo State, Brazil. *The Journal of Parasitology* 94: 1259–63. [CrossRef] [PubMed]

Rah, Hyungchul, Bruno B. Chomel, Erich H. Follmann, R. W. Kasten, C. H. Hew, Thomas Buseck Farver, Gerald W. Garner, and Steven C. Amstrup. 2005. Serosurvey of Selected Zoonotic Agents in Polar Bears (Ursus Maritimus). *The Veterinary Record* 156: 7–13. [CrossRef]

Rajkhowa, Swaraj, Dilip Kumar Sarma, and Chandan Rajkhowa. 2006. Seroprevalence of *Toxoplasma gondii* Antibodies in Captive Mithuns (Bos Frontalis) from India. *Veterinary Parasitology* 135: 369–74. [CrossRef] [PubMed]

Ramzan, M., M. Akhtar, F. Muhammad, I. Hussain, Elzbieta Hiszczyńska-Sawicka, A. U. Haq, M. S. Mahmood, and Mian Abdul Hafeez. 2009. Seroprevalence of *Toxoplasma gondii* in Sheep and Goats in Rahim Yar Khan (Punjab), Pakistan. *Tropical Animal Health and Production* 41: 1225–29. [CrossRef]

Razmi, Gholamreza R., V. Abedi, and S. Yaghfoori. 2016. Serological Study of *Toxoplasma gondii* Infection in Turkoman Horses in the North Khorasan Province, Iran. *Journal of Parasitic Diseases: Official Organ of the Indian Society for Parasitology* 40: 515–19. [CrossRef] [PubMed]

Remes, Noora, Age Kärssin, Kärt Must, Maarja Tagel, Brian Lassen, and Pikka Jokelainen. 2018. *Toxoplasma gondii* Seroprevalence in Free-Ranging Moose (Alces Alces) Hunted for Human Consumption in Estonia: Indicator Host Species for Environmental *Toxoplasma gondii* Oocyst Contamination. *Veterinary Parasitology, Regional Studies and Reports* 11: 6–11. [CrossRef] [PubMed]

Remington, Jack S., Philippe Thulliez, and Jose G. Montoya. 2004. Recent Developments for Diagnosis of Toxoplasmosis. *Journal of Clinical Microbiology* 42: 941–45. [CrossRef]

Rengifo-Herrera, Claudia, Edwin Pile, Anabel García, Alexander Pérez, Dimas Pérez, Felicia K. Nguyen, Valli de la Guardia, Rima Mcleod, and Zuleima Caballero. 2017. Seroprevalence of *Toxoplasma gondii* in Domestic Pets from Metropolitan Regions of Panama. *Parasite (Paris, France)* 24: 9. [CrossRef]

Rengifo-Herrera, Claudia, Luis Miguel Ortega-Mora, Gema Alvarez-García, Mercedes Gómez-Bautista, Daniel García-Párraga, Francisco Javier García-Peña, and Susana Pedraza-Díaz. 2012. Detection of *Toxoplasma gondii* Antibodies in Antarctic Pinnipeds. *Veterinary Parasitology* 190: 259–62. [CrossRef]

Richomme, Celine, Dominique Aubert, Emmanuelle Gilot-Fromont, Daniel Ajzenberg, Aurelien Mercier, Christian Ducrot, Hubert Ferté, Daniel Delorme, and Isabelle Villena. 2009. Genetic Characterization of *Toxoplasma gondii* from Wild Boar (Sus Scrofa) in France. *Veterinary Parasitology* 164: 296–300. [CrossRef]

Rissanen, Karoliina, Maryna Galat, Ganna Kovalenko, Olena Rodnina, Glib Mikharovskyi, Kärt Must, and Pikka Jokelainen. 2019. *Toxoplasma gondii* Seroprevalence in Horses from Ukraine: An Investigation Using Two Serological Methods. *Acta Parasitologica* 64: 687–92. [CrossRef]

Robert-Gangneux, Florence, Virginie Commere, Claudine Tourte-Schaefer, and J. Dupouy-Camet. 1999. Performance of a Western Blot Assay to Compare Mother and Newborn Anti-Toxoplasma Antibodies for the Early Neonatal Diagnosis of Congenital Toxoplasmosis. *European Journal of Clinical Microbiology & Infectious Diseases: Official Publication of the European Society of Clinical Microbiology* 18: 648–54. [CrossRef]

Roberts, A., Klaus Hedman, V. Luyasu, J. Zufferey, Marie Helene Bessières, R. M. Blatz, Ermanno Candolfi, A. Decoster, G. Enders, Uwe Groß, and et al. 2001. Multicenter Evaluation of Strategies for Serodiagnosis of Primary Infection with *Toxoplasma gondii*. *European Journal of Clinical Microbiology & Infectious Diseases: Official Publication of the European Society of Clinical Microbiology* 20: 467–74. [CrossRef]

Rodrigues, Filipa T., Fernando A. Moreira, Teresa Coutinho, Jitender P. Dubey, Luís Cardoso, and Ana Patrícia Lopes. 2019. Antibodies to *Toxoplasma gondii* in Slaughtered Free-Range and Broiler Chickens. *Veterinary Parasitology* 271: 51–53. [CrossRef]

Roghmann, Mary Claire, Charles T. Faulkner, A. Lefkowitz, Sharon Patton, Jeffrey James Zimmerman, and John Glenn Morris. 1999. Decreased Seroprevalence for *Toxoplasma gondii* in Seventh Day Adventists in Maryland. *The American Journal of Tropical Medicine and Hygiene* 60: 790–92. [CrossRef]

Roqueplo, Cedric, Radu Blaga, Marie Jean-Lou, Isabelle Vallee, and Bernard Davoust. 2017. Seroprevalence of *Toxoplasma gondii* in Hunted Wild Boars (Sus Scrofa) from Southeastern France. *Folia Parasitologica* 64: 003. [CrossRef]

Ross, Robin D., Lori A. Stec, Jane C. Werner, Mark S. Blumenkranz, Louis Glazer, and George A. Williams. 2001. Presumed Acquired Ocular Toxoplasmosis in Deer Hunters. *Retina (Philadelphia, PA)* 21: 226–29. [CrossRef]

Rouatbi, Mariem, Safa Amairia, Mustapha Lahmer, Narjess Lassoued, Mourad Rekik, Barbara Wieland, Joram M. Mwacharo, and Mohamed Gharbi. 2019. Detection of *Toxoplasma gondii* Infection in Semen of Rams Used for Natural Mating in Commercial Sheep Farms in Tunisia. *Veterinary Parasitology, Regional Studies and Reports* 18: 100341. [CrossRef]

Ryser-Degiorgis, Marie-Pierre, Eva-Britt Jakubek, Carl Hård af Segerstad, Caroline Bröjer, Torsten Mörner, Désirée S. Jansson, Anna Lundén, and Arvid Uggla. 2006. Serological Survey of *Toxoplasma gondii* Infection in Free-Ranging Eurasian Lynx (Lynx Lynx) from Sweden. *Journal of Wildlife Diseases* 42: 182–87. [CrossRef] [PubMed]

Saavedra, Geraldine M., and Ynés R. Ortega. 2004. Seroprevalence of *Toxoplasma gondii* in Swine from Slaughterhouses in Lima, Peru, and Georgia, U.S.A. *The Journal of Parasitology* 90: 902–4. [CrossRef]

Sacks, Jeffrey J., Ronald R. Roberto, and Norman F. Brooks. 1982. Toxoplasmosis Infection Associated with Raw Goat's Milk. *JAMA* 248: 1728–32. [CrossRef]

Sadek, O. A., Zeinab M. Abdel-Hameed, and Huda M. Kuraa. 2015. Molecular Detection of *Toxoplasma gondii* Dna In Raw Goat and Sheep Milk With Discussion Of Its Public Health Importance in Assiut Governorate. *Assiut Veterinary Medical Journal* 61: 166–77. [CrossRef]

Sadrebazzaz, Alireza, Hamidreza Haddadzadeh, and Parviz Shayan. 2006. Seroprevalence of Neospora Caninum and *Toxoplasma gondii* in Camels (Camelus Dromedarius) in Mashhad, Iran. *Parasitology Research* 98: 600–601. [CrossRef]

Sævik, Bente Kristin, Randi Ingebjørg Krontveit, Kristine P. Eggen, Nina Malmberg, Stein I. Thoresen, and Kristin W. Prestrud. 2015. *Toxoplasma gondii* Seroprevalence in Pet Cats in Norway and Risk Factors for Seropositivity. *Journal of Feline Medicine and Surgery* 17: 1049–56. [CrossRef] [PubMed]

Sah, Ramesh Prasad, Md Hasanuzzaman Talukder, A. K. M. Anisur Rahman, Mohammad Zahangir Alam, and Michael P. Ward. 2018. Seroprevalence of *Toxoplasma gondii* Infection in Ruminants in Selected Districts in Bangladesh. *Veterinary Parasitology, Regional Studies and Reports* 11: 1–5. [CrossRef] [PubMed]

Salant, H., T. Weingram, D. T. Spira, and T. Eizenberg. 2009a. An Outbreak of Toxoplasmosis amongst Squirrel Monkeys in an Israeli Monkey Colony. *Veterinary Parasitology* 159: 24–29. [CrossRef] [PubMed]

Salant, Harold, Alex Markovics, Dan T. Spira, and Joseph Hamburger. 2007. The Development of a Molecular Approach for Coprodiagnosis of *Toxoplasma gondii*. *Veterinary Parasitology* 146: 214–20. [CrossRef]

Salant, Harold, and Dan T. Spira. 2004. A Cross-Sectional Survey of Anti-*Toxoplasma gondii* Antibodies in Jerusalem Cats. *Veterinary Parasitology* 124: 167–77. [CrossRef]

Salant, Harold, Daniel Yasur Landau, and Gad Baneth. 2009b. A Cross-Sectional Survey of *Toxoplasma gondii* Antibodies in Israeli Pigeons. *Veterinary Parasitology* 165: 145–49. [CrossRef]

Sampaio, Barbara Fialho Carvalho, Miriam De Souza Macre, Luciana Regina Meireles, and Heitor Franco de Andrade Junior. 2014. Saliva as a Source of Anti-*Toxoplasma gondii* IgG for Enzyme Immunoassay in Human Samples. *Clinical Microbiology and Infection: The Official Publication of the European Society of Clinical Microbiology and Infectious Diseases* 20: O72–O74. [CrossRef]

Santos, Sara Lima, Kattyanne de Souza Costa, Leane Queiroz Gondim, Mariana Sampaio Anares da Silva, Rosângela Soares Uzêda, Kiyoko Abe-Sandes, and Luís Fernando Pita Gondim. 2010. Investigation of Neospora Caninum, *Hammondia* Sp., and *Toxoplasma gondii* in Tissues from Slaughtered Beef Cattle in Bahia, Brazil. *Parasitology Research* 106: 457–61. [CrossRef]

Santos, Thais Rabelo, Alvimar Jose da Costa, Gilson Helio Toniollo, Maria Cecilia Rui Luvizotto, Ana Helena Benetti, Raquel Ribeiro Santos, Danilo Henrique da Matta, Welber Daniel Zanetti Lopes, Joao Ademir de Oliveira, and Gilson Pereira de Oliveira. 2009. Prevalence of Anti-*Toxoplasma gondii* Antibodies in Dairy Cattle, Dogs, and Humans from the Jauru Micro-Region, Mato Grosso State, Brazil. *Veterinary Parasitology* 161: 324–26. [CrossRef]

Satbige, Ajay Suryakant, Mangalanathan Vijaya Bharathi, Parthiban I. Ganesan, Chirukandoth Sreekumar, and Chellaiah Rajendran. 2016. Detection of *Toxoplasma gondii* in Small Ruminants in Chennai Using PCR and Modified Direct Agglutination Test. *Journal of Parasitic Diseases: Official Organ of the Indian Society for Parasitology* 40: 1466–69. [CrossRef]

Sawadogo, P., Jamaleddine Hafid, Bahrie Bellete, Roger Tran Manh Sung, M. Chakdi, Pierre Flori, Helene Raberin, I. Bent Hamouni, Abderrahman Chait, and Abderrahim Dalal. 2005. Seroprevalence of *T. gondii* in Sheep from Marrakech, Morocco. *Veterinary Parasitology* 130: 89–92. [CrossRef] [PubMed]

Schale, Sarah, Daniel Howe, Michelle Yeargan, Jennifer K. Morrow, Amy Graves, and Amy L. Johnson. 2018. Protozoal Coinfection in Horses with Equine Protozoal Myeloencephalitis in the Eastern United States. *Journal of Veterinary Internal Medicine* 32: 1210–14. [CrossRef] [PubMed]

Schares, Gereon, Majda Globokar Vrhovec, Nikola Pantchev, Daland C. Herrmann, and Franz J. Conraths. 2008. Occurrence of *Toxoplasma gondii* and Hammondia Hammondi Oocysts in the Faeces of Cats from Germany and Other European Countries. *Veterinary Parasitology* 152: 34–45. [CrossRef] [PubMed]

Schares, Gereon, Nikola Pantchev, Dieter Barutzki, Alfred Otto Heydorn, Christian Bauer, and Franz J. Conraths. 2005. Oocysts of Neospora Caninum, Hammondia Heydorni, *Toxoplasma gondii* and Hammondia Hammondi in Faeces Collected from Dogs in Germany. *International Journal for Parasitology* 35: 1525–37. [CrossRef]

Schatzberg, Scott J., Nicholas J. Haley, Stephen C. Barr, Alexander deLahunta, Natasha Olby, Karen Munana, and Nicholas J. H. Sharp. 2003. Use of a Multiplex Polymerase Chain Reaction Assay in the Antemortem Diagnosis of Toxoplasmosis and Neosporosis in the Central Nervous System of Cats and Dogs. *American Journal of Veterinary Research* 64: 1507–13. [CrossRef]

Schulzig, Heike Silvia, and Karsten Fehlhaber. 2005. [Longitudinal study on the seroprevalence of Toxoplasma gondii infection in four German pig breeding and raising farms]. *Berliner Und Munchener Tierarztliche Wochenschrift* 118: 399–403.

Sechi, Paola, Antonio Ciampelli, Valentina Cambiotti, Fabrizia Veronesi, and Beniamino T. Cenci-Goga. 2013. Seroepidemiological Study of Toxoplasmosis in Sheep in Rural Areas of the Grosseto District, Tuscany, Italy. *Italian Journal of Animal Science* 12: e39. [CrossRef]

Sedlák, Kamil, Ivan Literák, Martin Faldyna, Miroslav Toman, and Jaromir Benák. 2000. Fatal Toxoplasmosis in Brown Hares (Lepus Europaeus): Possible Reasons of Their High Susceptibility to the Infection. *Veterinary Parasitology* 93: 13–28. [CrossRef]

Sepúlveda, Maximiliano A., Claudia Muñoz-Zanzi, Carla Rosenfeld, Rocio Jara, Katharine M. Pelican, and Dolores Hill. 2011. *Toxoplasma gondii* in Feral American Minks at the Maullín River, Chile. *Veterinary Parasitology* 175: 60–65. [CrossRef]

Sevgili, Murat, Cahit Babur, Ahmet Gokcen, and Serpil Nalbantoglu. 2004. Seroprevalence of *Toxoplasma gondii* (Nicolle and Manceaux, 1908) in Thoroughbred Arabian Mares as Detected by Sabin-Feldman Dye Test in Şanhurfa. *Veterinary Bulletin* 74: 1571.

Sevgili, Murat, Cahit Babür, Serpil Nalbantoğlu, Gülçin Karaş, and Zati Vatansever. 2005. Determination of Seropositivity for *Toxoplasma gondii* in Sheep in Şanlıurfa Province. *Turkish Journal of Veterinary & Animal Sciences* 29: 107–11.

Shaapan, Raafat M., Fathi A. El-Nawawi, and Mohamed A. A. Tawfik. 2008. Sensitivity and Specificity of Various Serological Tests for the Detection of *Toxoplasma gondii* Infection in Naturally Infected Sheep. *Veterinary Parasitology* 153: 359–62. [CrossRef] [PubMed]

Shakespeare, Martin. 2009. *Zoonoses*, 2nd ed. London: Pharmaceutical Press.

Sickinger, Eva, Hans-Bertram Braun, Gerald Praast, Myriam Stieler, Cordelia Gundlach, Claudia Birkenbach, and John Prostko. 2009. Evaluation of the Abbott ARCHITECT Toxo IgM Assay. *Diagnostic Microbiology and Infectious Disease* 64: 275–82. [CrossRef] [PubMed]

Silaghi, Cornelia, Martin Knaus, Dhimiter Rapti, Ilir Kusi, Enstela Shukullari, Dietmar Hamel, Kurt Pfister, and Steffen Rehbein. 2014. Survey of *Toxoplasma gondii* and Neospora Caninum, Haemotropic Mycoplasmas and Other Arthropod-Borne Pathogens in Cats from Albania. *Parasites & Vectors* 7: 62. [CrossRef]

Siloşi, I., S. Rogoz, A. Ungureanu, C. Siloşi, C. Avramescu, A. Muşetescu, O. Drackoulog, and C. Neamţiu. 2007. Investigaţii Imunoserologice În Toxoplasmoză. *Revista Română de Parazitologie* 17: 49–54.

Silva, Deise A. O., Janaína Lobato, Tiago W. P. Mineo, and José R. Mineo. 2007. Evaluation of Serological Tests for the Diagnosis of Neospora Caninum Infection in Dogs: Optimization of Cut off Titers and Inhibition Studies of Cross-Reactivity with *Toxoplasma gondii*. *Veterinary Parasitology* 143: 234–44. [CrossRef]

Silva, Jean Carlos Ramos, Solange Maria Gennari, Alessandra Mara Alves Ragozo, V. R. Amajones, C. Magnabosco, Lucia Eiko Oishi Yai, Jose Soares Ferreira-Neto, and Jitender P. Dubey. 2002. Prevalence of *Toxoplasma gondii* Antibodies in Sera of Domestic Cats from Guarulhos and São Paulo, Brazil. *The Journal of Parasitology* 88: 419–20. [CrossRef]

Silva, Jean Carlos, Saemi Ogassawara, Cristina Harumi Adania, Fernando Ferreira, Solange Maria Gennari, Jitender P. Dubey, and Jose Soares Ferreira-Neto. 2001. Seroprevalence of *Toxoplasma gondii* in Captive Neotropical Felids from Brazil. *Veterinary Parasitology* 102: 217–24. [CrossRef]

Silva, Karin Lucianne Monti de Vasconcellos, and Mário Luiz de La Rue. 2006. Possibilidade da transmissão congênita de Toxoplasma gondii em ovinos através de seguimento sorológico no município de Rosário do Sul, RS, Brasil. *Ciência Rural* 36: 892–97. [CrossRef]

Silva, Mariana S. A., Rosângela S. Uzêda, Kattyanne S. Costa, Sara L. Santos, Alan C. C. Macedo, Kiyoko Abe-Sandes, and Luis Fernando P. Gondim. 2009. Detection of Hammondia Heydorni and Related Coccidia (Neospora Caninum and *Toxoplasma gondii*) in Goats Slaughtered in Bahia, Brazil. *Veterinary Parasitology* 162: 156–59. [CrossRef]

Silva, R, C Bonassi, Osmar Dalla Costa, and Nelson Morés. 2003. Serosurvey on Toxoplasmosis in Outdoor Pig Production Systems of the Southern Region of Brazil. *Revue D Elevage Et De Medicine Veterinaire Des Pays Tropicaux* 56: 3–4.

Silva, Roberto Aguilar Machado Santos. 2005. Antibodies to *Toxoplasma gondii* in Horses from Pantanal, Brazil. *Veterinária e Zootecnia* 12: 20–24.

Slifko, Theresa R., Huw V. Smith, and Joan B. Rose. 2000. Emerging Parasite Zoonoses Associated with Water and Food. *International Journal for Parasitology* 30: 1379–93. [CrossRef] [PubMed]

Smielewska-Łoś, Ewa, and Waldemar Turniak. 2004. *Toxoplasma gondii* Infection in Polish Farmed Mink. *Veterinary Parasitology* 122: 201–6. [CrossRef]

Smith, Kirk E., John R. Fisher, and Jitender P. Dubey. 1995. Toxoplasmosis in a Bobcat (Felis Rufus). *Journal of Wildlife Diseases* 31: 555–57. [CrossRef]

Soares, Herbert S., Sílvia M. M. Ahid, Ana C. D. S. Bezerra, Hilda F. J. Pena, Ricardo A. Dias, and Solange M. Gennari. 2009. Prevalence of Anti-*Toxoplasma gondii* and Anti-Neospora Caninum Antibodies in Sheep from Mossoró, Rio Grande Do Norte, Brazil. *Veterinary Parasitology* 160: 211–14. [CrossRef]

Sobrino, Raquel, Oscar Cabezón, Javier Millán, Marcela Pabón, Maria Cruz Arnal, Daniel F. Luco, Christian Gortázar, Jitender P. Dubey, and Sonia Almeria. 2007. Seroprevalence of *Toxoplasma gondii* Antibodies in Wild Carnivores from Spain. *Veterinary Parasitology* 148: 187–92. [CrossRef] [PubMed]

Soiza, J. A., and H. Pastor. 1970. Toxoplasma Retinochoroiditis. *Archivos de oftalmologia de Buenos Aires* 45: 161–64.

Sørensen, Karen Kristine, Torill Mørk, Olof G. Sigurdardóttir, Kjetil Asbakk, Johan Akerstedt, Bjarne Asbjorn Bergsjø, and Eva Fuglei. 2005. Acute Toxoplasmosis in Three Wild Arctic Foxes (Alopex Lagopus) from Svalbard; One with Co-Infections of Salmonella Enteritidis PT1 and Yersinia Pseudotuberculosis Serotype 2b. *Research in Veterinary Science* 78: 161–67. [CrossRef] [PubMed]

Sotiriadou, Isaia, and Panagiotis Karanis. 2008. Evaluation of Loop-Mediated Isothermal Amplification for Detection of *Toxoplasma gondii* in Water Samples and Comparative Findings by Polymerase Chain Reaction and Immunofluorescence Test (IFT). *Diagnostic Microbiology and Infectious Disease* 62: 357–65. [CrossRef] [PubMed]

Spada, Eva, Daniela Proverbio, Alessandra Della Pepa, Giulia Domenichini, Giada Bagnagatti De Giorgi, Giorgio Traldi, and Elisabetta Ferro. 2013. Prevalence of Faecal-Borne Parasites in Colony Stray Cats in Northern Italy. *Journal of Feline Medicine and Surgery* 15: 672–77. [CrossRef] [PubMed]

Spada, Eva, Ilaria Canzi, Luciana Baggiani, Roberta Perego, Fabrizio Vitale, Antonella Migliazzo, and Daniela Proverbio. 2016. Prevalence of Leishmania Infantum and Co-Infections in Stray Cats in Northern Italy. *Comparative Immunology, Microbiology and Infectious Diseases* 45: 53–58. [CrossRef] [PubMed]

Speer, Clarence A., and Jitender P. Dubey. 2005. Ultrastructural Differentiation of *Toxoplasma gondii* Schizonts (Types B to E) and Gamonts in the Intestines of Cats Fed Bradyzoites. *International Journal for Parasitology* 35: 193–206. [CrossRef] [PubMed]

Spilovská, Silvia, Katarina Reiterová, Danka Kovácová, M. Bobáková, and Pavol Dubinský. 2009. The First Finding of Neospora Caninum and the Occurrence of Other Abortifacient Agents in Sheep in Slovakia. *Veterinary Parasitology* 164: 320–23. [CrossRef] [PubMed]

Spišák, František, Ľudmila Turčeková, Katarína Reiterová, Silvia Špilovská, and Pavol Dubinský. 2010. Prevalence Estimation and Genotypization of *Toxoplasma gondii* in Goats. *Biologia* 65: 670–74. [CrossRef]

Sposito Filha, E., V. do Amaral, R. Macruz, M.M. Reboucas, S.M. Santos, and F. Borgo. 1992. Infeccao Experimental de Equinos Com Taquizoitos de *Toxoplasma gondii*. *Revista Brasileira de Parasitologia Veterinária* 1: 51–54.

Sreekumar, Chirukandoth, Douglas H. Graham, Erica Dahl, Tovi Lehmann, Muthusamy Raman, Dilip P. Bhalerao, Manoella Campostrini Barreto Vianna, and Jitender P. Dubey. 2003. Genotyping of *Toxoplasma gondii* Isolates from Chickens from India. *Veterinary Parasitology* 118: 187–94. [CrossRef] [PubMed]

Srikitjakarn, Lertrak, W. Chaisaowong, G. Punyagosa, Anucha Sirimalaisuwan, and C. Saksangawong. 2008. Possibility to Certify *Toxoplasma gondii* Free Pig Production. *International Journal of Infectious Diseases* 12: e137. [CrossRef]

Sroka, Jacek, and Jolanta Szymańska. 2012. Analysis of Prevalence of *Toxoplasma gondii* Infection in Selected Rural Households in the Lublin Region. *Journal of Veterinary Research* 56: 529–34. [CrossRef]

Sroka, Jacek, Angelina Wojcik-Fatla, Jolanta Szymanska, Jacek Dutkiewicz, Violetta Zajac, and Jacek Zwolinski. 2010. The Occurrence of *Toxoplasma gondii* Infection in People and Animals from Rural Environment of Lublin Region—Estimate of Potential Role of Water as a Source of Infection. *Annals of Agricultural and Environmental Medicine: AAEM* 17: 125–32.

Sroka, Jacek, Jacek Karamon, Angelina Wójcik-Fatla, Weronika Piotrowska, Jacek Dutkiewicz, Ewa Bilska-Zając, Violetta Zając, Maciej Kochanowski, Joanna Dąbrowska, and Tomasz Cencek. 2020. *Toxoplasma gondii* Infection in Slaughtered Pigs and Cattle in Poland: Seroprevalence, Molecular Detection and Characterization of Parasites in Meat. *Parasites & Vectors* 13: 223. [CrossRef]

Sroka, Jacek, Jacek Karamon, Jacek Dutkiewicz, Angelina Wójcik Fatla, Violetta Zając, and Tomasz Cencek. 2018. Prevalence of *Toxoplasma gondii* Infection in Cats in Southwestern Poland. *Annals of Agricultural and Environmental Medicine: AAEM* 25: 576–80. [CrossRef]

Sroka, Jacek, Tomasz Cencek, Irena Ziomko, Jacek Karamon, and And Zwoliński. 2008. Preliminary Assessment of ELISA, MAT, and LAT for Detecting *Toxoplasma gondii* Antibodies in Pigs. *Bulletin- Veterinary Institute in Pulawy* 52: 545–49.

Stibbs, H. H. 1985. Changes in Brain Concentrations of Catecholamines and Indoleamines in *Toxoplasma gondii* Infected Mice. *Annals of Tropical Medicine and Parasitology* 79: 153–57. [CrossRef]

Stimbirys, Arturas, Jonas Bagdonas, Gediminas Gerulis, and Pierre Russo. 2007. A Serological Study on the Prevalence of *Toxoplasma gondii* in Sheep of Lithuania. *Polish Journal of Veterinary Sciences* 10: 83–87. [PubMed]

Ştirbu-Teofănescu, B., D. Militaru, I. Diaconu, and E. Bucur. 2007. Detecţia Parazitului *Toxoplasma gondii* În Probe Biologice Prin Imunofluorescenţă. *Revista Română de Parazitologie* 17: 176.

Stroehle, Angelika, Katja Schmid, Ivo Heinzer, Arunasalam Naguleswaran, and Andrew Hemphill. 2005. Performance of a Western Immunoblot Assay to Detect Specific Anti-*Toxoplasma gondii* IgG Antibodies in Human Saliva. *The Journal of Parasitology* 91: 561–63. [CrossRef]

Stroia, V., and A. Ungureanu. 2007. Toxoplasmoza Congenitală. *Revista Română de Parazitologie* 17: 77–80.

Styles, J., A. Egorov, R. Converse, E. Sams, and T. Wade. 2018. Evaluation of In-House Salivary IgG Antibody Assays for the Detection of Chronic *Toxoplasma gondii* and Helicobacter Pylori Infections. Available online: https://cfpub.epa.gov/si/si_public_record_report.cfm?Lab=NHEERL&dirEntryId=346048 (accessed on 18 November 2021).

Su, C., X. Zhang, and Jitender P. Dubey. 2006. Genotyping of *Toxoplasma gondii* by Multilocus PCR-RFLP Markers: A High Resolution and Simple Method for Identification of Parasites. *International Journal for Parasitology* 36: 841–48. [CrossRef] [PubMed]

Suaréz-Aranda, F., A. J. Galisteo, R. M. Hiramoto, R. P. Cardoso, L. R. Meireles, O. Miguel, and H. F. Andrade. 2000. The Prevalence and Avidity of *Toxoplasma gondii* IgG Antibodies in Pigs from Brazil and Peru. *Veterinary Parasitology* 91: 23–32. [CrossRef]

Sundar, N., R. A. Cole, N. J. Thomas, D. Majumdar, Jitender P. Dubey, and C. Su. 2008. Genetic Diversity among Sea Otter Isolates of *Toxoplasma gondii*. *Veterinary Parasitology* 151: 125–32. [CrossRef] [PubMed]

Szabo, K. A., M. G. Mense, T. P. Lipscomb, K. J. Felix, and Jitender P. Dubey. 2004. Fatal Toxoplasmosis in a Bald Eagle (Haliaeetus Leucocephalus). *The Journal of Parasitology* 90: 907–8. [CrossRef]

Tagel, Maarja, Brian Lassen, Arvo Viltrop, and Pikka Jokelainen. 2019. Large-Scale Epidemiological Study on *Toxoplasma gondii* Seroprevalence and Risk Factors in Sheep in Estonia: Age, Farm Location, and Breed Associated with Seropositivity. *Vector Borne and Zoonotic Diseases (Larchmont, N.Y.)* 19: 421–29. [CrossRef]

Talari, S.A., M. Arbabi, Z. Vakili, and H. Ahmadi. 2004. Comparison of Ig G, Ig M-IFA and Ig M-ELISA in *Toxoplasma gondii* Serology. *Journal of Zoology* 24: 33–36.

Tassi, P. 2007. *Toxoplasma gondii* Infection in Horses. A Review. *Parassitologia* 49: 7–15. [PubMed]

Tavalla, Mehdi, Fatemeh Asgarian, and Forough Kazemi. 2017. Prevalence and Genetic Diversity of *Toxoplasma gondii* Oocysts in Cats of Southwest of Iran. *Infection, Disease & Health* 22: 203–9. [CrossRef]

Teixeira, Weslen Fabricio Pires, Maria Eduarda Gonçalves Tozato, Giovana Pavão Vital, Welber Daniel Zanetti Lopes, Samea Fernandes Joaquim, Amanda Leal de Vasconcellos, Flávia Carolina Fávero, Lívia Mendonça Pascoal, Gustavo Felippelli, Helio Langoni, and et al. 2020. Investigação parasitológica e molecular de Toxoplasma gondii em urina e saliva de felinos (Felis catus) infectados experimentalmente. *Research, Society and Development* 9: e923975143. [CrossRef]

Tekay, Fikret, and Erdal Ozbek. 2007. [The seroprevalence of Toxoplasma gondii in women from Sanliurfa, a province with a high raw meatball consumption]. *Turkiye Parazitolojii Dergisi* 31: 176–79. [PubMed]

Tekkesin, Nilgun, Kenan Keskin, Cumhur Kılınc, Nuray Orgen, and Kaya Molo. 2011. Detection of Immunoglobulin G Antibodies to *Toxoplasma gondii*: Evaluation of Two Commercial Immunoassay Systems. *Journal of Microbiology, Immunology, and Infection = Wei Mian Yu Gan Ran Za Zhi* 44: 21–26. [CrossRef]

Tenter, Astrid M., Anja R. Heckeroth, and Louis M. Weiss. 2000. *Toxoplasma gondii*: From Animals to Humans. *International Journal for Parasitology* 30: 1217–58. [PubMed]

Teodorescu, C., V. Manea, and I. Teodorescu. 2007. Asocierea Metodelor PCR Şi ELISA Pentru Detectarea Infestării Femeilor Gravide Cu *Toxoplasma gondii*. *Revista Română de Parazitologie* 17: 37–42.

Teshale, S., A. Dumètre, M. L. Dardé, B. Merga, and P. Dorchies. 2007. Serological Survey of Caprine Toxoplasmosis in Ethiopia: Prevalence and Risk Factors. *Parasite (Paris, France)* 14: 155–59. [CrossRef] [PubMed]

Thakur, Rashmi, Rajnish Sharma, R. S. Aulakh, J. P. S. Gill, and B. B. Singh. 2019. Prevalence, Molecular Detection and Risk Factors Investigation for the Occurrence of *Toxoplasma gondii* in Slaughter Pigs in North India. *BMC Veterinary Research* 15: 431. [CrossRef] [PubMed]

Thaw, Yu Nandi, Tin Aye Khaing, Kyaw San Linn, Soe Soe Wai, Lat Lat Htun, and Saw Bawm. 2021. The First Seroepidemiological Study on *Toxoplasma gondii* in Backyard Pigs in Myanmar. *Parasite Epidemiology and Control* 14: e00216. [CrossRef]

Thiangtum, Khongsak, Burin Nimsuphun, Nongnuch Pinyopanuwat, Wissanuwat Chimnoi, Wanchai Tunwattana, Daraka Tongthainan, Sathaporn Jittapalapong, Theera Rukkwamsuk, and Soichi Maruyama. 2006. Seroprevalence of *Toxoplasma gondii* in Captive Felids in Thailand. *Veterinary Parasitology* 136: 351–55. [CrossRef]

Thiele, D. 1990. The Technique of Polymerase Chain Reaction--a New Diagnostic Tool in Microbiology and Other Scientific Fields (Review). *Zentralblatt Fur Bakteriologie: International Journal of Medical Microbiology* 273: 431–54. [CrossRef] [PubMed]

Thiptara, Anyarat, Wandee Kongkaew, Umai Bilmad, Teeraphun Bhumibhamon, and Somsak Anan. 2006. Toxoplasmosis in Piglets. *Annals of the New York Academy of Sciences* 1081: 336–38. [CrossRef] [PubMed]

Tilahun, Berhanu, Yacob Hailu Tolossa, Getachew Tilahun, Hagos Ashenafi, and Shihun Shimelis. 2018. Seroprevalence and Risk Factors of *Toxoplasma gondii* Infection among Domestic Ruminants in East Hararghe Zone of Oromia Region, Ethiopia. *Veterinary Medicine International* 2018: 4263470. [CrossRef]

Titilincu, Adriana, Viorica Mircean, Radu Blaga, Lidia Chiţimia, and Vasile Cozma. 2008. Cercetări Privind Prevalenţa Infecţiei Cu *Toxoplasma gondii* La Pisici. *Revista Română de Medicină Veterinară* 2: 108–22.

Toulah, Fawzia H., Saedia A. Sayed Al-Ahl, Dawlat M. Amin, and Mona H. Hamouda. 2011. *Toxoplasma gondii*: Ultrastructure Study of the Entry of Tachyzoites into Mammalian Cells. *Saudi Journal of Biological Sciences* 18: 151–56. [CrossRef]

Tsutsui, V.S., I.T. Navarro, R.L. Freire, J.C. Freitas, L.B. Prdencio, A.C.B. Delbem, and E.R.M. Marana. 2004. Serum Epidemiology and Associated Factors on Swine Transmission of *Toxoplasma gondii* at Northern Parana, Brazil. *Veterinary Bulletin* 74: 933.

Tuda, Josef, Sri Adiani, Madoka Ichikawa-Seki, Kousuke Umeda, and Yoshifumi Nishikawa. 2017. Seroprevalence of *Toxoplasma gondii* in Humans and Pigs in North Sulawesi, Indonesia. *Parasitology International* 66: 615–18. [CrossRef] [PubMed]

Tuntasuvan, D., K. Mohkaew, and J. P. Dubey. 2001. Seroprevalence of *Toxoplasma gondii* in Elephants (Elephus Maximus Indicus) in Thailand. *The Journal of Parasitology* 87: 229–30. [CrossRef] [PubMed]

Turčeková, Ľudmila, Daniela Antolová, Katarína Reiterová, and František Spišák. 2013. Occurrence and Genetic Characterization of *Toxoplasma gondii* in Naturally Infected Pigs. *Acta Parasitologica* 58: 361–66. [CrossRef]

Turner, C. B., and D. Savva. 1990. Evidence of *Toxoplasma gondii* in an Equine Placenta. *The Veterinary Record* 127: 96.

Turner, C. B., and D. Savva. 1991. Detection of *Toxoplasma gondii* in Equine Eyes. *The Veterinary Record* 129: 128. [CrossRef] [PubMed]

Turner, C. B., and D. Savva. 1992. Transplacental Infection of a Foal with *Toxoplasma gondii*. *The Veterinary Record* 131: 179–80. [CrossRef]

Tzanidakis, Nikolaos, Pavlo Maksimov, Franz J. Conraths, Evaggelos Kiossis, Christos Brozos, Smaragda Sotiraki, and Gereon Schares. 2012. *Toxoplasma gondii* in Sheep and Goats: Seroprevalence and Potential Risk Factors under Dairy Husbandry Practices. *Veterinary Parasitology* 190: 340–48. [CrossRef]

Ueno, Tatiana Evelyn Hayama, Vitor Salvador Picão Gonçalves, Marcos Bryan Heinemann, Tales Luís Bezerra Dilli, Bruno Minoru Akimoto, Silvio Luís Pereira de Souza, Solange Maria Gennari, and Rodrigo Martins Soares. 2009. Prevalence of *Toxoplasma gondii* and Neospora Caninum Infections in Sheep from Federal District, Central Region of Brazil. *Tropical Animal Health and Production* 41: 547–52. [CrossRef]

Uggla, A., S. Mattson, and N. Juntti. 1990. Prevalence of Antibodies to *Toxoplasma gondii* in Cats, Dogs and Horses in Sweden. *Acta Veterinaria Scandinavica* 31: 219–22. [CrossRef] [PubMed]

Ullmann, Leila Sabrina, Rodrigo Costa da Silva, Wanderlei de Moraes, Zalmir Silvino Cubas, Leonilda Correia dos Santos, Juliano Leônidas Hoffmann, Nei Moreira, Ana Marcia Sa Guimaraes, Patrícia Montaño, Helio Langoni, and et al. 2010. Serological Survey of *Toxoplasma gondii* in Captive Neotropical Felids from Southern Brazil. *Veterinary Parasitology* 172: 144–46. [CrossRef]

van de Velde, Norbert, Brecht Devleesschauwer, Mardik Leopold, Lineke Begeman, Lonneke IJsseldijk, Sjoukje Hiemstra, Jooske IJzer, Andrew Brownlow, Nicholas Davison, Jan Haelters, and et al. 2016. *Toxoplasma gondii* in Stranded Marine Mammals from the North Sea and Eastern Atlantic Ocean: Findings and Diagnostic Difficulties. *Veterinary Parasitology* 230: 25–32. [CrossRef]

Van der Giessen, Joke, Manoj Fonville, Martijn Bouwknegt, Merel Langelaar, and Ant Vollema. 2007. Seroprevalence of *Trichinella spiralis* and *Toxoplasma gondii* in Pigs from Different Housing Systems in The Netherlands. *Veterinary Parasitology* 148: 371–74. [CrossRef]

van Knapen, F., A. F. Kremers, J. H. Franchimont, and U. Narucka. 1995. Prevalence of Antibodies to *Toxoplasma gondii* in Cattle and Swine in The Netherlands: Towards an Integrated Control of Livestock Production. *The Veterinary Quarterly* 17: 87–91. [CrossRef]

Venturini, M. C., D. Bacigalupe, L. Venturini, M. Machuca, C. J. Perfumo, and Jitender P. Dubey. 1999. Detection of Antibodies to *Toxoplasma gondii* in Stillborn Piglets in Argentina. *Veterinary Parasitology* 85: 331–34. [CrossRef]

Venturini, Maria Cecilia, Diana Bacigalupe, Lucila Venturini, Magdalena Rambeaud, Walter Basso, Juan M. Unzaga, and Carlos J. Perfumo. 2004. Seroprevalence of *Toxoplasma gondii* in Sows from Slaughterhouses and in Pigs from an Indoor and an Outdoor Farm in Argentina. *Veterinary Parasitology* 124: 161–65. [CrossRef]

Verdeş, (Hotea) Ionela. 2011. The Seroprevalence of *Toxoplasma gondii* Infection in Animals, from Western Romania. Ph.D. thesis, Banat University of Agricultural Sciences and Veterinary Medicine Timisoara, Facultaty of Veterinary Medicine, Timisoara, Romania.

Verin, Ranieri, Linda Mugnaini, Simona Nardoni, Roberto Amerigo Papini, Gaetano Ariti, Alessandro Poli, and Francesca Mancianti. 2013. Serologic, Molecular, and Pathologic Survey of *Toxoplasma gondii* Infection in Free-Ranging Red Foxes (Vulpes Vulpes) in Central Italy. *Journal of Wildlife Diseases* 49: 545–51. [CrossRef]

Verma, Shiv K., Larissa Minicucci, Darby Murphy, Michelle Carstensen, Carolin Humpal, Paul Wolf, Rafael Calero-Bernal, Camila K. Cerqueira-Cézar, Oliver C. H. Kwok, Chunlei Su, and et al. 2016. Antibody Detection and Molecular Characterization of *Toxoplasma gondii* from Bobcats (Lynx Rufus), Domestic Cats (Felis Catus), and Wildlife from Minnesota, USA. *The Journal of Eukaryotic Microbiology* 63: 567–71. [CrossRef]

Veronesi, Fabrizia, Azzurra Santoro, Giovanni L. Milardi, Manuela Diaferia, Giulia Morganti, David Ranucci, and Simona Gabrielli. 2017. Detection of *Toxoplasma gondii* in Faeces of Privately Owned Cats Using Two PCR Assays Targeting the B1 Gene and the 529-Bp Repetitive Element. *Parasitology Research* 116: 1063–69. [CrossRef]

Vesco, G., W. Buffolano, S. La Chiusa, G. Mancuso, S. Caracappa, A. Chianca, S. Villari, V. Currò, F. Liga, and E. Petersen. 2007. *Toxoplasma gondii* Infections in Sheep in Sicily, Southern Italy. *Veterinary Parasitology* 146: 3–8. [CrossRef] [PubMed]

Villari, S., G. Vesco, E. Petersen, A. Crispo, and W. Buffolano. 2009. Risk Factors for Toxoplasmosis in Pigs Bred in Sicily, Southern Italy. *Veterinary Parasitology* 161: 1–8. [CrossRef]

Vitaliano, S. N., D. A. O. Silva, T. W. P. Mineo, R. A. Ferreira, E. Bevilacqua, and J. R. Mineo. 2004. Seroprevalence of *Toxoplasma gondii* and Neospora Caninum in Captive Maned Wolves (Chrysocyon Brachyurus) from Southeastern and Midwestern Regions of Brazil. *Veterinary Parasitology* 122: 253–60. [CrossRef]

Vollaire, Melissa R., Steven V. Radecki, and Michael R. Lappin. 2005. Seroprevalence of *Toxoplasma gondii* Antibodies in Clinically Ill Cats in the United States. *American Journal of Veterinary Research* 66: 874–77. [CrossRef] [PubMed]

Vostalová, E., Ivan Literák, I. Pavlásek, and Kamil Sedlak. 2000. Prevalence of *Toxoplasma gondii* in Finishing Pigs in a Large-Scale Farm in the Czech Republic. *Acta Veterinaria Brno* 69: 209–12. [CrossRef]

Vyas, Ajai, Seon-Kyeong Kim, Nicholas Giacomini, John C. Boothroyd, and Robert M. Sapolsky. 2007. Behavioral Changes Induced by Toxoplasma Infection of Rodents Are Highly Specific to Aversion of Cat Odors. *Proceedings of the National Academy of Sciences of the United States of America* 104: 6442–47. [CrossRef]

Waap, Helga, Anabela Vilares, Eugénia Rebelo, Salomé Gomes, and Helena Angelo. 2008. Epidemiological and Genetic Characterization of *Toxoplasma gondii* in Urban Pigeons from the Area of Lisbon (Portugal). *Veterinary Parasitology* 157: 306–9. [CrossRef]

Waap, Helga, Rita Cardoso, Alexandre Leitão, Telmo Nunes, Anabela Vilares, Maria João Gargaté, José Meireles, Helder Cortes, and Helena Ângelo. 2012. In Vitro Isolation and Seroprevalence of *Toxoplasma gondii* in Stray Cats and Pigeons in Lisbon, Portugal. *Veterinary Parasitology* 187: 542–47. [CrossRef] [PubMed]

Waap, Helga, Telmo Nunes, Yolanda Vaz, and Alexandre Leitão. 2016. Serological Survey of *Toxoplasma gondii* and Besnoitia Besnoiti in a Wildlife Conservation Area in Southern Portugal. *Veterinary Parasitology, Regional Studies and Reports* 3–4: 7–12. [CrossRef] [PubMed]

Wallace, G. D. 1969. Serologic and Epidemiologic Observations on Toxoplasmosis on Three Pacific Atolls. *American Journal of Epidemiology* 90: 103–11. [CrossRef] [PubMed]

Wallander, C., J. Frössling, I. Vågsholm, A. Uggla, and A. Lundén. 2015. *Toxoplasma gondii* Seroprevalence in Wild Boars (Sus Scrofa) in Sweden and Evaluation of ELISA Test Performance. *Epidemiology and Infection* 143: 1913–21. [CrossRef]

Wallander, Camilla, Jenny Frössling, Fernanda C. Dórea, Arvid Uggla, Ivar Vågsholm, and Anna Lundén. 2016. Pasture Is a Risk Factor for *Toxoplasma gondii* Infection in Fattening Pigs. *Veterinary Parasitology* 224: 27–32. [CrossRef]

Wang, H.-L., G.-H. Wang, Q.-Y. Li, C. Shu, M.-S. Jiang, and Y. Guo. 2006. Prevalence of Toxoplasma Infection in First-Episode Schizophrenia and Comparison between Toxoplasma-Seropositive and Toxoplasma-Seronegative Schizophrenia. *Acta Psychiatrica Scandinavica* 114: 40–48. [CrossRef]

Wanha, K., R. Edelhofer, C. Gabler-Eduardo, and H. Prosl. 2005. Prevalence of Antibodies against Neospora Caninum and *Toxoplasma gondii* in Dogs and Foxes in Austria. *Veterinary Parasitology* 128: 189–93. [CrossRef] [PubMed]

Webster, J. P. 1994. Prevalence and Transmission of *Toxoplasma gondii* in Wild Brown Rats, Rattus Norvegicus. *Parasitology* 108, Pt 4: 407–11. [CrossRef] [PubMed]

Webster, J. P. 2001. Rats, Cats, People and Parasites: The Impact of Latent Toxoplasmosis on Behaviour. *Microbes and Infection* 3: 1037–45. [CrossRef] [PubMed]

Webster, J. P., P. H. L. Lamberton, C. A. Donnelly, and E. F. Torrey. 2006. Parasites as Causative Agents of Human Affective Disorders? The Impact of Anti-Psychotic, Mood-Stabilizer and Anti-Parasite Medication on *Toxoplasma gondii*'s Ability to Alter Host Behaviour. *Proceedings. Biological Sciences* 273: 1023–30. [CrossRef] [PubMed]

Webster, Joanne P. 2010. Dubey, Jitender P. Toxoplasmosis of Animals and Humans. *Parasites & Vectors* 3: 112. [CrossRef]

Webster, Joanne P., and Glenn A. McConkey. 2010. *Toxoplasma gondii*-Altered Host Behaviour: Clues as to Mechanism of Action. *Folia Parasitologica* 57: 95–104. [CrossRef]

Weigel, R. M., Jitender P. Dubey, A. M. Siegel, D. Hoefling, D. Reynolds, L. Herr, U. D. Kitron, S. K. Shen, P. Thulliez, and R. Fayer. 1995. Prevalence of Antibodies to *Toxoplasma gondii* in Swine in Illinois in 1992. *Journal of the American Veterinary Medical Association* 206: 1747–51.

Weigel, R. M., Jitender P. Dubey, D. Dyer, and A. M. Siegel. 1999. Risk Factors for Infection with *Toxoplasma gondii* for Residents and Workers on Swine Farms in Illinois. *The American Journal of Tropical Medicine and Hygiene* 60: 793–98. [CrossRef]

Weiss, J. B. 1995. DNA Probes and PCR for Diagnosis of Parasitic Infections. *Clinical Microbiology Reviews* 8: 113–30. [CrossRef]

Weiss, Louis M., and Kami Kim. 2004. *Toxoplasma gondii*: The Model Apicomplexan. *International Journal for Parasitology* 34: 423–32. [CrossRef]

White, Thomas J. 1996. The Future of PRC Technology: Diversification of Technologies and Applications. *Trends in Biotechnology* 14: 478–83. [CrossRef] [PubMed]

Wilson, C. B., J. S. Remington, S. Stagno, and D. W. Reynolds. 1980. Development of Adverse Sequelae in Children Born with Subclinical Congenital Toxoplasma Infection. *Pediatrics* 66: 767–74. [CrossRef] [PubMed]

Winiecka-Krusnell, Jadwiga, Isabel Dellacasa-Lindberg, J. P. Dubey, and Antonio Barragan. 2009. *Toxoplasma gondii*: Uptake and Survival of Oocysts in Free-Living Amoebae. *Experimental Parasitology* 121: 124–31. [CrossRef] [PubMed]

Wolf, D., G. Schares, O. Cardenas, W. Huanca, Aida Cordero, Andrea Bärwald, F. J. Conraths, M. Gauly, H. Zahner, and C. Bauer. 2005. Detection of Specific Antibodies to Neospora Caninum and *Toxoplasma gondii* in Naturally Infected Alpacas (Lama Pacos), Llamas (Lama Glama) and Vicunas (Lama Vicugna) from Peru and Germany. *Veterinary Parasitology* 130: 81–87. [CrossRef]

Wyss, R., H. Sager, N. Müller, F. Inderbitzin, M. König, L. Audigé, and B. Gottstein. 2000. [The occurrence of Toxoplasma gondii and Neospora caninum as regards meat hygiene]. *Schweizer Archiv Fur Tierheilkunde* 142: 95–108.

Xiao, Jianchun, Stephen L. Buka, Tyrone D. Cannon, Yasuhiro Suzuki, Raphael P. Viscidi, E. Fuller Torrey, and Robert H. Yolken. 2009. Serological Pattern Consistent with Infection with Type I *Toxoplasma gondii* in Mothers and Risk of Psychosis among Adult Offspring. *Microbes and Infection* 11: 1011–18. [CrossRef] [PubMed]

Xie, DeHua, Xingquan Zhu, Huanlin Cui, Chuji Qui, Wanhong Fan, Shenquan Liao, Minli Zhai, Ruiquing Lin, and Yabiao Weng. 2005. Development of PCR Assay for Diagnosing Swine Toxoplasmosis. *Chinese Journal of Veterinary Science and Technology* 35: 289–93.

Xing, Hai, Lixin Xu, Xiaokai Song, Xiangrui Li, and Ruofeng Yan. 2018. Seroprevalence of *Toxoplasma gondii* and *Trichinella spiralis* in Horses in Xinjiang, Northwestern China. *Journal of Equine Veterinary Science* 60: 11–15. [CrossRef]

Yabsley, Michael J., Carly N. Jordan, Sheila M. Mitchell, Terry M. Norton, and David S. Lindsay. 2007. Seroprevalence of *Toxoplasma gondii*, Sarcocystis Neurona, and Encephalitozoon Cuniculi in Three Species of Lemurs from St. Catherines Island, GA, USA. *Veterinary Parasitology* 144: 28–32. [CrossRef]

Yagmur, Fatih, Suleyman Yazar, Hanife Ozcan Temel, and Mustafa Cavusoglu. 2010. May *Toxoplasma gondii* Increase Suicide Attempt-Preliminary Results in Turkish Subjects? *Forensic Science International* 199: 15–17. [CrossRef] [PubMed]

Yai, Lúcia E. O., Alessandra M. A. Ragozo, Daniel M. Aguiar, José T. Damaceno, Luciana N. Oliveira, J. P. Dubey, and Solange M. Gennari. 2008. Isolation of *Toxoplasma gondii* from Capybaras (Hydrochaeris Hydrochaeris) from São Paulo State, Brazil. *The Journal of Parasitology* 94: 1060–63. [CrossRef]

Yan, C., C. L. Yue, S. B. Qiu, H. L. Li, H. Zhang, H. Q. Song, S. Y. Huang, F. C. Zou, M. Liao, and X. Q. Zhu. 2011. Seroprevalence of *Toxoplasma gondii* Infection in Domestic Pigeons (Columba Livia) in Guangdong Province of Southern China. *Veterinary Parasitology* 177: 371–73. [CrossRef]

Yan, C., C. L. Yue, Z. G. Yuan, Y. He, C. C. Yin, R. Q. Lin, Jitender P. Dubey, and X. Q. Zhu. 2009. *Toxoplasma gondii* Infection in Domestic Ducks, Free-Range and Caged Chickens in Southern China. *Veterinary Parasitology* 165: 337–40. [CrossRef]

Yang, Yurong, Yuqing Ying, S. K. Verma, A. B. M. Cassinelli, O. C. H. Kwok, Hongde Liang, A. K. Pradhan, X. Q. Zhu, C. Su, and Jitender P. Dubey. 2015. Isolation and Genetic Characterization of Viable *Toxoplasma gondii* from Tissues and Feces of Cats from the Central Region of China. *Veterinary Parasitology* 211: 283–88. [CrossRef]

Yarim, Gul Fatma, Cevat Nisbet, Taraneh Oncel, Sena Cenesiz, and Gulay Ciftci. 2007. Serum Protein Alterations in Dogs Naturally Infected with *Toxoplasma gondii*. *Parasitology Research* 101: 1197–202. [CrossRef] [PubMed]

Ybañez, Rochelle Haidee D., Chadinne Girlani R. Busmeon, Alexa Renee G. Viernes, Jorim Z. Langbid, Johanne P. Nuevarez, Adrian P. Ybañez, and Yoshifumi Nishikawa. 2019. Endemicity of Toxoplasma Infection and Its Associated Risk Factors in Cebu, Philippines. *PLoS ONE* 14: e0217989. [CrossRef] [PubMed]

Yekkour, Feriel, Dominique Aubert, Aurélien Mercier, Jean-Benjamin Murat, Mammar Khames, Paul Nguewa, Khatima Ait-Oudhia, Isabelle Villena, and Zahida Bouchene. 2017. First Genetic Characterization of *Toxoplasma gondii* in Stray Cats from Algeria. *Veterinary Parasitology* 239: 31–36. [CrossRef] [PubMed]

Yereli, Kor, I. Cüneyt Balcioğlu, and Ahmet Ozbilgin. 2006. Is *Toxoplasma gondii* a Potential Risk for Traffic Accidents in Turkey? *Forensic Science International* 163: 34–37. [CrossRef]

Yildiz, Kader, Oguz Kul, Cahit Babur, Selcuk Kilic, Aycan N. Gazyagci, Bekir Celebi, and I. Safa Gurcan. 2009. Seroprevalence of Neospora Caninum in Dairy Cattle Ranches with High Abortion Rate: Special Emphasis to Serologic Co-Existence with *Toxoplasma gondii*, Brucella Abortus and Listeria Monocytogenes. *Veterinary Parasitology* 164: 306–10. [CrossRef]

Yu, Jinhai, Jun Ding, Zhaofei Xia, Degui Lin, Yili Li, Junying Jia, and Qun Liu. 2008. Seroepidemiology of *Toxoplasma gondii* in Pet Dogs and Cats in Beijing, China. *Acta Parasitologica* 53: 317–19. [CrossRef]

Yu, Jinhai, Zhaofei Xia, Qun Liu, Jing Liu, Jun Ding, and Wei Zhang. 2007. Seroepidemiology of Neospora Caninum and *Toxoplasma gondii* in Cattle and Water Buffaloes (Bubalus Bubalis) in the People's Republic of China. *Veterinary Parasitology* 143: 79–85. [CrossRef]

Yu, XiuLing, Nanhua Chen, Dongmei Hu, Wei Zhang, Xiaoxia Li, Baoyue Wang, Liping Kang, Xiangdong Li, Qun Liu, and Kegong Tian. 2009. Detection of Neospora Caninum from Farm-Bred Young Blue Foxes (Alopex Lagopus) in China. *The Journal of Veterinary Medical Science* 71: 113–15. [CrossRef]

Yücesan, Banuçiçek, Cahit Babür, Nafiye Koç, Figen Sezen, Selçuk Kılıç, and Yüksel Gürüz. 2019. Investigation of Anti-*Toxoplasma gondii* Antibodies in Cats Using Sabin-Feldman Dye Test in Ankara in 2016. *Turkiye Parazitolojii Dergisi* 43: 5–9. [CrossRef]

Zarnke, R. L., Jitender P. Dubey, O. C. Kwok, and J. M. Ver Hoef. 1997. Serologic Survey for *Toxoplasma gondii* in Grizzly Bears from Alaska. *Journal of Wildlife Diseases* 33: 267–70. [CrossRef]

Zarra-Nezhad, Fatemeh, Mahdi P. Borujeni, Bahman Mosallanejad, and Hossein Hamidinejat. 2017. A Seroepidemiological Survey of *Toxoplasma gondii* Infection in Referred Dogs to Veterinary Hospital of Ahvaz, Iran. *International Journal of Veterinary Science and Medicine* 5: 148–51. [CrossRef]

Zhang, Houshuang, Oriel M. M. Thekisoe, Gabriel O. Aboge, Hisako Kyan, Junya Yamagishi, Noboru Inoue, Yoshifumi Nishikawa, Satoshi Zakimi, and Xuenan Xuan. 2009. *Toxoplasma gondii*: Sensitive and Rapid Detection of Infection by Loop-Mediated Isothermal Amplification (LAMP) Method. *Experimental Parasitology* 122: 47–50. [CrossRef]

Zhang, S. Y., S. F. Jiang, Y. Y. He, C. E. Pan, M. Zhu, and M. X. Wei. 2004. Serologic Prevalence of *Toxoplasma gondii* in Field Mice, Microtus Fortis, from Yuanjiang, Hunan Province, People's Republic of China. *The Journal of Parasitology* 90: 437–38. [CrossRef]

Zhang, Yehua, Haiyan Gong, Rongsheng Mi, Yan Huang, Xiangan Han, Luming Xia, Shoufu Li, Haiyan Jia, Xiaoli Zhang, Tao Sun, and et al. 2020. Seroprevalence of *Toxoplasma gondii* Infection in Slaughter Pigs in Shanghai, China. *Parasitology International* 76: 102094. [CrossRef]

Zhou, Mo, Shinuo Cao, Ferda Sevinc, Mutlu Sevinc, Onur Ceylan, Mingming Liu, and Guanbo Wang. 2017. Enzyme-Linked Immunosorbent Assays Using Recombinant TgSAG2 and Nc-SAG1 to Detect *Toxoplasma gondii* and Neospora Caninum-Specific Antibodies in Domestic Animals in Turkey. *The Journal of Veterinary Medical Science* 78: 1877–81. [CrossRef]

Zhu, Jibao, Jigang Yin, Yue Xiao, Ning Jiang, Johan Ankarlev, Johan Lindh, and Qijun Chen. 2008. A Sero-Epidemiological Survey of *Toxoplasma gondii* Infection in Free-Range and Caged Chickens in Northeast China. *Veterinary Parasitology* 158: 360–63. [CrossRef]

Zimmerman, J. J., D. W. Dreesen, W. J. Owen, and G. W. Beran. 1990. Prevalence of Toxoplasmosis in Swine from Iowa. *Journal of the American Veterinary Medical Association* 196: 266–70.

Zintl, A., D. Halova, G. Mulcahy, J. O'Donovan, B. Markey, and T. DeWaal. 2009. In Vitro Culture Combined with Quantitative TaqMan PCR for the Assessment of *Toxoplasma gondii* Tissue Cyst Viability. *Veterinary Parasitology* 164: 167–72. [CrossRef]

Zou, Feng-Cai, Xiu-Tao Sun, Yin-Jie Xie, Bing Li, Guo-Hong Zhao, Gang Duan, and Xing-Quan Zhu. 2009. Seroprevalence of *Toxoplasma gondii* in Pigs in Southwestern China. *Parasitology International* 58: 306–7. [CrossRef]

MDPI
St. Alban-Anlage 66
4052 Basel
Switzerland
www.mdpi.com

MDPI Books Editorial Office
E-mail: books@mdpi.com
www.mdpi.com/books